Springer Series in
OPTICAL SCIENCES 97

Founded by H.K.V. Lotsch

Editor in Chief: W.T. Rhodes, Atlanta

Editorial Board: T. Asakura, Sapporo
K.-H. Brenner, Mannheim
T.W. Hänsch, Garching
T. Kamiya, Tokyo
F. Krausz, Vienna and Garching
B. Monemar, Linköping
H. Venghaus, Berlin
H. Weber, Berlin
H. Weinfurter, Munich

Springer
New York
Berlin
Heidelberg
Hong Kong
London
Milan
Paris
Tokyo

Physics and Astronomy

ONLINE LIBRARY

springeronline.com

Springer Series in
OPTICAL SCIENCES

The Springer Series in Optical Sciences, under the leadership of Editor-in-Chief *William T. Rhodes*, Georgia Institute of Technology, USA, and Georgia Tech Lorraine, France, provides an expanding selection of research monographs in all major areas of optics: lasers and quantum optics, ultrafast phenomena, optical spectroscopy techniques, optoelectronics, quantum information, information optics, applied laser technology, industrial applications, and other topics of contemporary interest.

With this broad coverage of topics, the series is of use to all research scientists and engineers who need up-to-date reference books.

The editors encourage prospective authors to correspond with them in advance of submitting a manuscript. Submission of manuscripts should be made to the Editor-in-Chief or one of the Editors. See also http://www.springer.de/phys/books/optical_science/

Editor-in-Chief
William T. Rhodes
Georgia Institute of Technology
School of Electrical and Computer Engineering
Atlanta, GA 30332-0250, USA
E-mail: bill.rhodes@ece.gatech.edu

Editorial Board
Toshimitsu Asakura
Hokkai-Gakuen University
Faculty of Engineering
1-1, Minami-26, Nishi 11, Chuo-ku
Sapporo, Hokkaido 064-0926, Japan
E-mail: asakura@eli.hokkai-s-u.ac.jp

Karl-Heinz Brenner
Chair of Optoelectronics
University of Mannheim
Institute of Computer Engineering
B6, 26
68131 Mannheim, Germany
E-mail: brenner@uni-mannheim.de

Theodor W. Hänsch
Max-Planck-Institut für Quantenoptik
Hans-Kopfermann-Strasse 1
85748 Garching, Germany
E-mail: t.w.haensch@physik.uni-muenchen.de

Takeshi Kamiya
Ministry of Education, Culture, Sports
Science and Technology
National Institution for Academic Degrees
3-29-1 Otsuka, Bunkyo-ku
Tokyo 112-0012, Japan
E-mail: kamiyatk@niad.ac.jp

Ferenc Krausz
Vienna University of Technology
Photonics Institute
Gusshausstrasse 27/387
1040 Wien, Austria
E-mail: ferenc.krausz@tuwien.ac.at
and
Max-Planck-Institut für Quantenoptik
Hans-Kopfermann-Strasse 1
85748 Garching, Germany

Bo Monemar
Department of Physics
and Measurement Technology
Materials Science Division
Linköping University
58183 Linköping, Sweden
E-mail: bom@ifm.liu.se

Herbert Venghaus
Heinrich-Hertz-Institut
für Nachrichtentechnik Berlin GmbH
Einsteinufer 37
10587 Berlin, Germany
E-mail: venghaus@hhi.de

Horst Weber
Technische Universität Berlin
Optisches Institut
Strasse des 17. Juni 135
10623 Berlin, Germany
E-mail: weber@physik.tu-berlin.de

Harald Weinfurter
Ludwig-Maximilians-Universität München
Sektion Physik
Schellingstrasse 4/III
80799 München, Germany
E-mail: harald.weinfurter@physik.uni-muenchen.de

Jürgen Jahns   Karl-Heinz Brenner
Editors

Microoptics
From Technology to Applications

With 292 Illustrations

Foreword by G. Michael Morris

 Springer

Jürgen Jahns
Optische Nachrichtentechnik
Universität Hagen
D-58084 Hagen
Germany

Karl-Heinz Brenner
LS Optoelektronik
Universität Mannheim
68131 Mannheim
Germany

Sep/æ
phys

Library of Congress Cataloging-in-Publication Data
Jahns, Jurgen, 1953–
    Microoptics : from technology to applications / Jürgen Jahns, K.-H. Brenner.
      p. cm. — (Springer series in optical sciences ; v. 2700)
    Includes bibliographical references and index.
    ISBN 0-387-20980-8 (alk. paper)
     1. Integrated optics.   2. Optoelectronic devices.   3. Miniature electronic equipment.   I.
Brenner, K.-H. (Karl Heinz) II. Title. III. Series.
    TA1660.J34   2004
    621.36′93—dc22   2004041722

ISBN 0-387-20980-8   Printed on acid-free paper.

Printed in the United States of America.   (MVY)

9 8 7 6 5 4 3 2 1   SPIN 10949078

Springer-Verlag is a part of *Springer Science+Business Media*

springeronline.com

Foreword

Microoptics: From Technology to Applications

In optics and photonics, like in many areas of science and technology, the frontier for the field lies in the constant push to reduce the size of components, devices, and systems. Since the early 1980s, the advances made in the design, fabrication, and application of microoptic components, devices, and systems have been truly astounding.

Leveraging on the revolution in fabrication technology in the electronics industry, optical scientists and engineers have perfected optical and electron-beam lithography as tools to write complex (binary and continuous) structured surfaces into photoresist with unprecedented precision. This precision-structured photoresist surface serves as a "master" element, which can be used to create hard or soft tools that are suitable for replication or transfer into a variety of materials. For example, a thin layer of gold may be deposited onto the photoresist, which, in turn, serves as an electrode in a nickel-plating process. The resulting nickel tool can then be used to replicate the elements into a variety of plastics using compression- or injection-molding techniques; or the tool could be used to replicate the surface using a "cast and cure" technique, in which a substrate, typically glass, is coated with a thin layer of ultraviolet- or temperature-cured polymer. In this case, the tool is pressed into the thin polymer, and the polymer is cured. Polymer-on-glass components are more stable, with respect to temperature variations, than pure plastic components. It is also possible to etch the precision-structured surface directly in a substrate material, such as fused silica, silicon, or germanium, using reactive-ion etching techniques. This can be accomplished either by etching structured photoresist surface into the original substrate or by first replicating the photoresist master element into an "etch" resist coated on a different substrate, which is subsequently transferred into the substrate via reactive-ion etching.

In addition to the advances in lithographic fabrication of microoptical components, there have been substantial improvements in the development of several other fabrication methods as well, such as single-point diamond turning, the LIGA process, and the use of pulsed and high-power laser systems to shape surfaces.

With the advances in fabrication technology to produce precision optical surfaces came the need for new and improved performance modeling methods of the microcomponents. In many cases, conventional scalar diffraction theory cannot adequately describe the operation of many of the components, spawning the application of rigorous electromagnetic modeling of the optical elements. Sophisticated computer models were developed to describe accurately diffraction efficiencies of diffractive structures and the polarization properties of both diffractive and high-numerical-aperture refractive microoptic components, and diffractive optics and waveguide modeling methods began to merge in order to model the performance of integrated microoptic systems.

Finally, there have been numerous advancements in microoptic metrology techniques and methods. After all, you cannot improve a process unless you can measure it! Significant advances have been made in virtually every area of microoptics metrology: contact surfaces probes, noncontact techniques for surface profiling, wave-front measurements of microoptic components, and microoptic system performance analysis. Also, there have been significant advances in new techniques to achieve micron and submicron alignment tolerances between microoptic components, thereby enabling the fabrication and application of novel microoptic devices and systems.

Over the past couple of decades, the microoptics field has made great strides in the transition to laboratory (or proof-of-principle) demonstration to high-volume commercial applications. Hats off to the scientists and engineering teams around the world that made this stunning success story a reality!

Rochester
September 2003

G. Michael Morris
RPC Photonics, Inc. & Apollo Optical Systems, LLC

Preface

Jürgen Jahns[1] and Karl-Heinz Brenner[2]

[1] FernUniversität Hagen, Optische Nachrichtentechnik, Universitätsstr. 27/PRG,
58084 Hagen, Germany
`juergen.jahns@fernuni-hagen.de`
[2] Universität Mannheim, Lehrstuhl für Optoelektronik, B6, 23-29, Bauteil C,
Zi.: 3.5, 68131 Mannheim, Germany
`brenner@uni-mannheim.de`

Optics is a key technology for many areas of applications. The famous study
"Harnessing Light" [1] defines the role of optics as that of a "pervasive en-
abler," which means that optical components can often be found to be the
decisive building block of a much larger technical area. One specific example
is the enormous development of the communications world after the advent
of the laser and fiber optics in the 1960s and 1970s. For the present and the
future, many other areas of application can be identified, for example, in information technology (transport, processing and storage of information), health
care and life sciences, sensing, lighting, manufacturing, and so forth. The study
also distinctly points out the interrelation between the progress in these areas
of application and the underlying progress of the optical technology in terms
of materials, devices, and systems.

Microoptics is one of those technologies for which the above-mentioned
considerations are becoming obvious. Similar to microelectronics, where component miniaturization and integration has opened a wide field of applications
and enabled continuing progress for over several decades, the miniaturized integration in other areas, such as optoelectronics, optics, and mechanics, may
provide a similar potential. Due to advanced fabrication techniques that have
been developed over the years, microoptical and micromechanical components
have now become feasible. Silicon micromechanical systems, arrays of microlenses, and vertical cavity surface emitting lasers (VCSEL) are among the
most prominent examples for this progress. At the same time, the demand for
practical optomechanical systems with improved functionality has increased.
The combination of microelectrical and micromechanical components in one
system (MEMS) has become a major area of research and development. By
integrating microoptical, microelectrical, and micromechanical components
(MOEMS) in one system, an even wider range of applications, including optical sensors, communication subsystems, and so forth, becomes possible.

The field of microoptics includes both, waveguide optics and free-space optics, although they differ in their mechanisms for light propagation and practical use. Optical signals offer a large degree of functionality based on the properties of electromagnetic radiation. The high bandwidth is utilized in telecommunications. The transfer of information or energy without physical contact is used most prominently in optical data storage and in optical material processing. Other advantages, such as the interferometric precision of coherent optics, are being exploited, for example, in sensors.

During recent years, the development of microoptics was strongly influenced by advances in technology and components. Although there is still progress in this area, systems aspects are becoming more important. Issues like modularization and integration of heterogeneous technologies (e.g., semiconductor, glass, plastic) are gaining increased significance. Another important issue is the definition of standardized interfaces, allowing one to freely combine different components or subsystems.

Consequently, new aspects of basic science are introduced. These include, for example, the wave-optical design of miniaturized systems, thermal aspects of hybrid integration, or the development of new techniques for characterizing microcomponents and systems. Also it must be pointed out, that a change of paradigm has occurred during the past decade: The emphasis in research and development has gradually shifted from component optimiation to system optimiation, taking into account manufacturability, practicality, lifetime, and cost. Due to this, new topics, such as packaging, tolerant system design, and so forth, are becoming more important.

Lithographic fabrication forms the foundation for practically all areas of microoptics. We distinguish between mask-based lithography adapted from the fabrication of integrated circuits (IC) and direct-write techniques using optical or electron beams. Conventional mask-based lithography uses optical illumination in the visible or near ultraviolet (UV). This technology is very suitable to implement, for example, diffractive microoptical components with quantized, staircaselike phase profile. Unlike in electronics, where the structure heights are typically in the nanometer range, optical structures often require much deeper profiles. For diffractive elements, a structural height that is on the order of a few wavelengths is required. For refractive elements and optomechanical structures, profile depths of up to several hundred microns are needed. For the realiation of deep structures, optical lithography with special resists can be used. Very high aspect ratios can be achieved with the LIGA (Lithographie, Galvanik, Abformung) approach, also discussed in this book. Another technological challenge is the fabrication of lateral structures with feature sizes in the subwavelength range, including the so-called "zero-order gratings" and the photonic crystals. Finally, another difficult challenge is the fabrication of 3-D structures. Unlike periodic 3-D structures, which can be implemented by interferometric techniques, irregular 3-D structures require a novel approach. The direct-write technique using ultrashort laser pulses for material structuring is currently under investigation.

Several of the new techniques just mentioned are described in this book. Kley, Wittig, and Tünnermann present and demonstrate the various tasks and techniques for lithographic fabrication. Conventional binary lithography is discussed at the beginning of the chapter and then the various forms of analog or gray-level lithography. The specific aspects of pattern generation for lithographic fabrication are discussed by Schnabel, with an emphasis on data economics, particularly for direct writing.

After this general introduction to lithographic fabrication in the first two chapters, we continue with the aspects of modeling and simulation of micro-optics. For free-space optical elements and systems, the different approaches are described in the chapter by Zeitner, Schreiber, and Karthe. Ray-optical and wave-optical approaches are discussed and the required modifications when modeling elements with a small diameter and/or very fine features. The next two chapters are devoted to the modeling of waveguide optics. First, März gives a general overview of the use of computer-aided engineering in optics and describes the different modeling approaches, like eigenvalue analysis, time-domain, and frequency-domain techniques. Then, Pregla describes the Method of Lines, a semianalytical technique, and shows its specific time-saving properties for periodic structures like Bragg gratings.

After this theoretical excursion, we step back to technology and turn toward the LIGA method as an approach to make very deep structures. The chapters by Frese and by Mohr, Last, and Wallrabe illuminate this interesting technology and its use for microoptics. Frese introduces the technology and presents waveguide-optical and free-space optical components as examples. The capability of the LIGA approach to fabricate very deep structures makes it interesting not only for the implementation of optical functions but also for micromechanics and packaging. The latter aspect is in the focus of the chapter by Mohr, Last, and Wallrabe, who present a modular approach to design and fabrication with the LIGA technology.

Microoptics has established itself over a relatively short time span due to the successful adaptation of a number of fabrication technologies. However, even more fabrication concepts have emerged over the past couple of years. A block of three chapters is devoted to such novel concepts. We begin with the use of fs-laser pulses as a tool for a direct-write approach. This has become possible with the development of practical fs-laser systems. The approach is of interest because it allows one to generate "arbitrary" 3-D structures (e.g., waveguides in a solid block of glass or plastic), as it is described by Nolte, Will, Burghoff, and Tünnermann. In the next chapter, Brinkmann, Reichel, and Hayden consider the possibilities that are further opened up by adding new bonding approaches and doping to the aforementioned fs-pulse direct-write technique. Finally, this block is completed by the paper by Jamois, Wehrspohn, Hermann, Hess, Andreani, and Gösele on 2-D waveguide structures based on the concept of photonic crystals. They consider so-called "high-Δn structures" made of semiconductor materials. Special room is given to the modeling and simulation of the light propagation in these devices.

An important aspect that is often neglected is the characterization of microoptics. Obviously, given the wide variety of microoptical components, this is a huge field which will gain increasing importance as commercial use increases. Here, Lindlein, Lamprecht and Schwider discuss the testing of microlenses. Several interferometric techniques are described which allow one to measure different parameters for different types of elements. An interesting fact worthwhile to mention is that some of these classical interferometric techniques have been adapted by including microoptical elements as optical "standards".

The next block of chapters deals with the topic of systems integration (in this aspect, related, of course, to the two LIGA chapters). Oikawa presents the fabrication of planar gradient-index microlenses and their use in building modules and systems, for example, for telecom applications. Gruber and Jahns follow with a chapter on the planar integration of free-space optical systems using thick slabs of transparent materials. One special purpose of this approach is the integration of 3-D optical interconnects for data communications in computers. Finally, Späth describes not only novel semiconductor light sources and their use for microoptics but also, in particular, the use of microoptical concepts for their efficient implementation.

The last topic of this book deals with applications, mostly in optical sensing. A general overview of microoptical sensors is given by Brandenburg. The main perspective of this work is, of course, how to use microfabrication to build miniaturized and integrated sensors. Typical examples are described, from distance measurement to spectroscopy. An example of the successful conversion into a industrial product is described by Rasch, Handrich, Spahlinger, Hafen, Voigt, and Weingärtner. They describe the use of LiNbO$_3$ waveguide technology for fiber-optic gyroscopes. The use of spectral encoding as one of many degrees of freedom in the design of sensor systems is discussed by Bartelt in his contribution. The book is concluded by Grunwald and Kebbel. Their work aims at the tremendous variety of microoptical components and their use in the shaping of ultrashort laser pulses.

This book understands itself as a current overview of the state of the art and as a complement the existing literature [2–6]. It comprises in written and updated form many presentations given at a Heraeus seminar on Microoptics held in Bad Honnef, Germany in April 2002. The editors would like to acknowledge the work of the authors who took precious time to write the manuscripts. Furthermore, the collaboration with Dr. Rasch (LITEF) and Dr. Ross (IOTech and M.u.T.) during the preparation of the Heraeus seminar is gratefully acknowledged. Thanks goes to the Heraeus foundation for its support and for providing a rather unique "ambiente." In the preparation of this book, the diligence and competence of Susanne Conradi and Tina Heldt (University of Hagen) was invaluable. Finally, we would like to thank Margaret Mitchell and

Hans Koelsch from Springer-Verlag, New York for their encouragement and patience.

Hagen and Mannheim   *Jürgen Jahns and Karl-Heinz Brenner*

November 2003

References

1. Committee on Optical Science and Engineering, *Harnessing Light: Optical Science and Engineering for the 21st Century*, National Academy Press, Washington, D.C. (1998).
2. Iga, K., Kokubun, Y., and Oikawa, M., *Fundamentals of Microoptics*, Academic Press, Tokyo (1984).
3. Herzig, H.-P. (ed.), *Micro-optics: Elements, Systems and Applications*, Taylor & Francis, London (1997).
4. Turunen, J. and Wyrowski, F. (eds.), *Diffractive Optics for Industrial and Commercial Applications*, Akademie Verlag, Berlin (1997).
5. Kufner, M. and Kufner, S., *Micro-optics and Lithography*, VUB Press, Brussels (1997)
6. Sinzinger, S. and Jahns, J., *Microoptics*, 2nd ed., Wiley-VCH, Weinheim (2003).

Contents

Microstructure Technology for Optical Component Fabrication

E.-Bernhard Kley, Lars-Christian Wittig, and Andreas Tünnermann

Institute of Applied Physics, Friedrich-Schiller-University Jena, Max-Wien-Platz 1, 07743 Jena, Germany
kley@iap.uni-jena.de

1 Introduction

Technologies and fabrication methods are often responsible for the success of products, because its price, functionality, size, and durability depend on it. Usually, technological reasons may lead to a time lag of years or even tens of years between the discovery and the first proof or the first commercial use of a physical effect. Therefore, the technologies play a key role in our technical life and we need to focus also on the fabrication technologies in a general discussion on microoptical elements.

Although microoptics shares the technological basis with microelectronics, the structuring techiques may differ considerably. This is caused by the fact that the microstructures that are necessary for the functionality of the optical elements can roughly be divided into different classes. The requirements of the profile shape are very different in the different classes and the specifications and tolerances cannot be generalized for the structures of microoptical elements. For this reason, it is necessary to understand the classification of microstructures and the optical properties that are related. Of course, it is not possible to refer to all fabrication methods in this chapter; even the lithographic methods we want to describe are of a huge variety that cannot be discussed completely. Therefore, an overview of the most important lithographic methods for optical element fabrication will be given and some basic knowledge of microlithography as well as equipment is expected.

2 Elements to Be Fabricated

The term "microoptical elements" contains a nearly infinite variety of different surface profiles. However, with respect to the different optical functions, we can classify the elements and define requirements for the corresponding fabrication technologies. Such a classification is, of course, not sharp and does not take into account the full variety of existing optical elements.

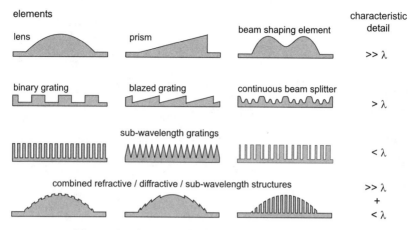

Fig. 1. Profiles of optical elements to be fabricated.

The first class consists of elements with continuous surface profiles such as lenses and prisms or, more general, beam-shaping elements with arbitrary shapes (Fig. 1, top row). The characteristic detail is large compared to the wavelength of light, and in order to analyze the optical function of these elements, the geometrical optical approach is sufficient. Ray tracing and the application of the law of refraction is the basis for modeling and design of these elements and that is why they are called refractive. However, with decreasing diameter or increasing deflection angle, wave-optical effects will affect the optical function more and more and have to be considered. Although from a physical point of view, optical elements are scalable with the wavelength, the size of these elements is mostly determined by technical demands of the setup or the device that admits the element. Thus, the aperture of these elements ranges from several microns to millimeters and the value for the total profile depth also varies over a wide range ($< 1\,\mu$m to $1\,$mm). The requirements for the profile accuracy for this class of elements are scaling with the wavelength and can be found usually between $\lambda/5$ and $\lambda/20$ ($2000\,$nm at $\lambda = 10\,\mu$m to $5\,$nm at $\lambda = 100\,$nm).

A second class of elements consists of periodic structures like binary and blazed gratings as well as beam splitters with a continuous profile (Fig. 1, second row). Computer-generated holograms (often used synonyms are kinoforms, diffractive optical elements, and holographic optical elements) also belong to this class. The characteristic details of these elements are larger than the wavelength of light, and for the analysis of its optical function, wave-optical methods have to be applied. Because diffraction of light is the basis for the functionality, these elements are called diffractive. In contrast to refractive elements, the accuracy of the transmitted (or reflected) wave front is related to the accuracy of the period d. If, for instance, the positioning error of a grating line is $d/10$, then the wave-front error will be $\lambda/10$ in the 1st and -1st diffraction orders. The accuracy of the profile shape mainly influences the distribution of intensity between different diffraction orders. The size of diffractive elements ranges from some micrometers to some hundred millimeters, but the profile depth scales with the wavelength. In the case of binary elements in the paraxial domain and suppressed zero-order efficiency, the profile depth is given by $(k + 1/2)(\lambda/\Delta n)$ (with k being an integer and Δn the difference of refractive indexes of the substrate and the surrounding medium), which corresponds to a phase step of $(2k + 1)\pi$. Typical values are $k = 1$ and $\Delta n = 0.457$ (fused silica at $\lambda = 633\,\text{nm}$), leading to a profile depth of nearly 700 nm.

The third class of elements we want to distinguish consists of periodic structures too, but the feature size is, in contrast to the above-mentioned diffractive elements, lower than the wavelength of light. Therefore, for normal incidence, no propagating diffraction orders exist and the structure acts as an effective medium. The term "artificial materials" is often used for this class of elements, because they have optical properties (e.g., birefringence, locally varying refractive index, polarization) which are very different from the original properties of the used material. Examples for elements are phase-retarder plates, moth eye structures, or blazed binary gratings (Fig. 1, third row). For the analysis of these elements (especially the calculation of the wave front just behind the element), rigorous methods have to be applied. Since there exists up to now no algorithm which allows the backward calculation of a profile from a given wave front, the design of these elements has to utilize a systematic scanning of the parameter space. The accuracy of the optical function scales with the wavelength and depends mainly on the accuracy of the grating parameters period d, fill factor f, and grating depth D. In some cases, d, f, and D have to be as accurate as 1%, which means a tolerance of only a few nanometers for typical values of a $\lambda/4$-phase retarder ($d = 400\,\text{nm}$, $f = 0.5$, $D = 2000\,\text{nm}$). The aperture of such elements lies in the range from several micrometers to some hundred millimeters, as in the case of diffractive elements.

Finally and most challenging, there is also the combination of all three classes of microoptical elements. This can be, for instance, a hybrid refractive/diffractive element or a refractive element made of an artificial material (Fig. 1, bottom row). In these cases, the requirements to accuracy of both the primary profile with characteristic detail $> \lambda$ as well as the secondary profile with characteristic detail $< \lambda$ has to be fulfilled. The size and profile depth of the complex structures are typically the same as for the primary class of elements.

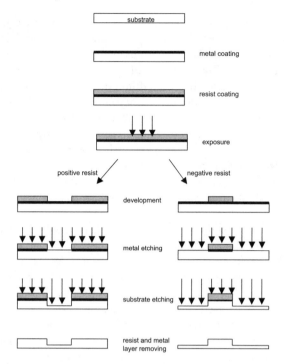

Fig. 2. Flowchart for the fabrication of binary optical elements (schematically).

3 Technologies

3.1 Fabrication Method for Binary Patterns

A binary optical pattern seems to be the easiest case for fabrication. As a matter of fact, the techniques for its fabrication are the standard lithographic fabrication technique sketched in Figure 2. In a first step, resist has to be coated on the substrate. The technique of choice depends on parameters like the resist thickness, the substrate size, and the shape of the surface or substrate, respectively. Spin coating is most common for coating standard resists on standard wafers or mask blanks. In this case, the resist thickness can be controlled by the substrate rotation speed, which usually lies in the range from 1000 to 6000 rpm and leads to a thickness variation of a factor 2 to 3 or a bit more. Resist thicknesses from 10 nm up to more than 100 μm can be made by using commercial resists. The next step is the resist exposure by light (photolithography), electrons (electron lithography), X-rays (X-ray lithography), or ions (ion lithography).

The "cheapest" piece of equipment for photolithography is a mask aligner for contact or proximity lithography. It uses a uniform beam of ultraviolet (UV) light (e.g., 365 nm or 315 nm) to illuminate a mask (typically made of chromium). Behind the mask, the resist-coated wafer is placed. With this shadow projection, minimum linewidths down to about 1.0 to 0.5 µm can be achieved, usually on substrates with a size of up to 6 in. and even larger. Better resolution can be achieved with the much more expensive projection photolithography (using so-called wafer steppers), which uses an objective for a reduction imaging of the mask into the focus or wafer plane. With advanced systems, a minimum linewidth below 100 nm could be demonstrated. Because of the reduction, the projected area covers only a part of the wafer. The full wafer exposure therefore works stepwise with the help of a very precise x/y-stage for wafer positioning. Advanced projection systems are able to work on a 12-in. wafer size.

For the pattern generation, this means direct writing or mask writing; the common methods are laser-beam writing or electron-beam (e-beam) writing. Laser writers are working on substrates with a size of up to 12 in. and generate linewidths down to about 400 nm at ultrawide (UV) wavelengths and about 200 nm for deep ultraviolet (DUV) wavelengths. Its address grids in the range between 50 nm and 1 nm. Similar parameters are known for the e-beam writers. However, the minimum linewidth often reaches values below 100 nm. Extreme values of the minimum linewidth are between 10 nm and 1 nm. A more detailed consideration of the pattern generation with laser writers and e-beam machines is given in Chapter 2.

Schematic sketches of the equipment described are shown in Figure 3.

Upon developement, a positive or negative copy of the mask pattern appears in the resist layer. With a subsequent etching step, it is possible to transfer the resist pattern into the substrate or into the layer below the resist. An improved etching mask can be made by using a chromium layer between the resist and the substrate. This layer has to be etched after the development process by dry etching. The chromium mask acts as a stable etch mask with high etching resistance in a reactive quartz etching process, for instance, and helps to achieve an excellent edge quality. Even though binary pattern fabrication techniques are extremely well developed for microelectronics, optical element fabrication may be a challenge due to special demands:

- Low pattern redundancies (e.g., Fourier Computer Generated Holograms (CGHs))
- Extreme aspect ratios and high accuracy (e.g., $\lambda/4$ effective media or photonic crystals)
- Continuous bended smooth pattern/gratings (e.g., waveguides or imaging gratings)
- High-resolution pattern on bended or topological surfaces (e.g., antireflection pattern on lenses)

It is clear that in order to overcome the problems resulting from these demands, new technologies have to be developed. However, it does not make sense to improve the fabrication methods blindly for fulfilling the specs or the tolerancing. Usually, the pushing demands are a result of optical designs without consideration of the fabrication techniques and often solutions can be

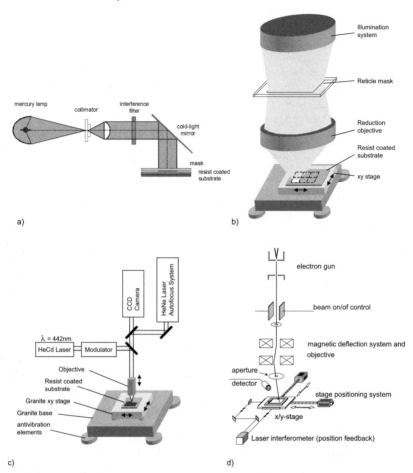

Fig. 3. Schematic sketches of the lithography equipment: (a) mask aligner, (b) wafer stepper, (c) laser writer, and (d) e-beam writer.

found by collaboration between designer and process engineer. Close interaction between the two can be beneficial – in particular, in cases where demands are high. This can be demonstrated by examples.

1. Diffractive optical elements (DOEs) with a large deflection angle are based on small pixel sizes, for instance. Figure 4 shows an example with 400 nm × 400 nm pixel size. Electron-beam writing with a variable shape beam should be a suitable and fast fabrication method. The advantage of variable shape beam writing is that the square-shaped pixels of the DOE can be written by the square-shaped e-beam. Hence, a short writing time of the DOE is guaranteed. However, because of the small pixel size of 400 nm × 400 nm, a bad influence on the geometrical pattern quality by the well-known proximity effect has to be expected. Indeed, a look at the

simulation of the proximity effect shows a rounded pattern (Fig. 4). Due to the large pixelated field without redundancies, a proximity-effect correction would lead to a substructure that increases the number of shapes to be written enormously. As a consequence, the writing time increases proportionally. In reality, the proximity effect does not disturb the optical function of this element. Suitable modeling shows an improved optical signal and an increased signal-to-noise ratio. In this case, the proximity effect makes the pattern worse but improves the optical properties of the element and a proximity correction would not make sense.

2. Grating-based phase plates (e.g., for $\lambda/4$-phase retardation) have to be considered as an alternative to classical phase plates based on anisotropic crystals. This means, however, that an accuracy of the phase retardation of about 1% has to be achieved. The geometrical specification of the relevant grating design shows for fused silica and 633-nm wavelength in transition the following:

Grating period	400 nm
Grating profile	rectangular
Fill factor	50% ± 1%
Grating depth	2200 nm ± 22 nm

Any deviation from a rectangular grating profile is not taken into account actually. Thus, the fabrication of such gratings seems to be impossible with such an accuracy. One way out of this dilemma comes with the understanding of the working principle of the grating and with the knowledge about the fabrication possibilities. Due to the small grating period, the grating is a zero-order grating and acts as an effective medium. Hence, the phase retardation between transverse electric (TE) and transverse magnetic (TM) polarization depends linearly on the birefringence that is controlled by the fill factor (includes the profile deviation) and on the grating depth. Inaccuracies of fill factor and profile deviations can be totally compensated by an adapted grating depth. With this knowledge, trimming methods can be used in practice for getting the high accuracy of phase retardation wanted. The first method is a back trimming. If the phase retardation of the grating fabricated is too large, a degradation of the grating profile by ion-beam etching under an angle reduces the phase retardation. The second method is suitable for increasing the phase retardation by increasing the grating depth. In this case, an etching mask has to be placed on the grating by the shadow technique and a further dry-etching step increases the grating depth (Fig. 5). After mask removing and phase retardation measured this process can be repeated a couple of times if necessary [1].

A similar problem is the fabrication of zero-order metal stripe gratings for polarization purposes in transmission. The aim is a high polarization ratio and so the design results in inconvenient grating specs. For a polar-

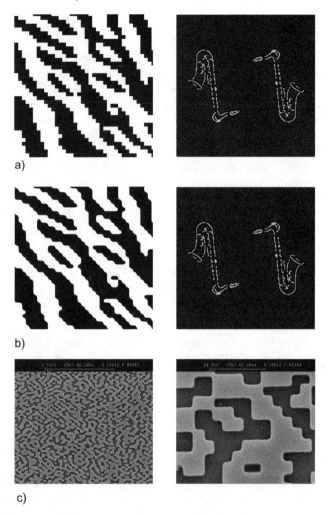

Fig. 4. Diffractive optical element with 400 nm × 400 nm pixel size: (a) detail of the ideal pattern and optical signal (theory); (b) detail of the pattern with proximity effect and optical signal (theory); the signal quality is not decreased by the proximity effect but increased slightly; (c) details of the pattern fabricated in fused silica.

ization ratio larger than 100 in the visible-wavelength range, we have two alternative designs:

	version A	*version B*
Grating period	100 nm	300 nm
Fill factor	50%	50%
Chromium stripe thickness	100 nm	500 nm

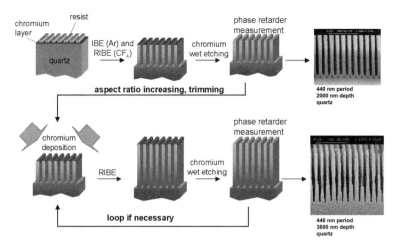

Fig. 5. Fabrication and trimming of high-aspect-ratio gratings.

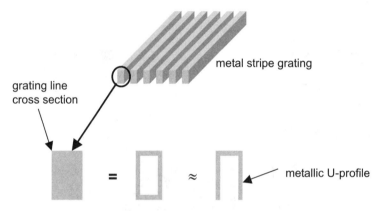

Fig. 6. Because of the skin effect, the massive metallic grating line cross section can be replaced by a hollow conductor or a u-profile as well.

Both variants are challenging for the fabrication. The small grating period of 100 nm is the problem of variant A; the high aspect ratio resulting from 500-nm chromium stripe thickness is the problem of variant B.

As we know, because light is a electromagnetic wave with a very high frequency and does penetrate metals just in the 10-nm range (skin effect), it is not necessary to fabricate massive metallic grating lines. The surface region only contributes to the absorption necessary for the polarization filter. Therefore, the massive grating line should be replaceable by profiles like shown in Figure 6. The modeling of a u-profile confirms this assumption. For the u-profile fabrication, binary grating fabrication and the shadow technique can be used (Fig. 7).

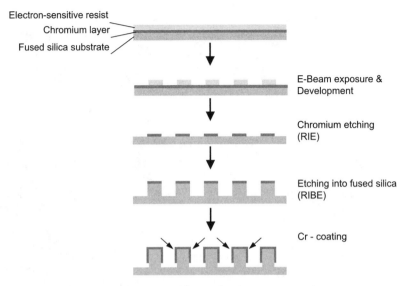

Fig. 7. Flowchart of the u-profile fabrication technique.

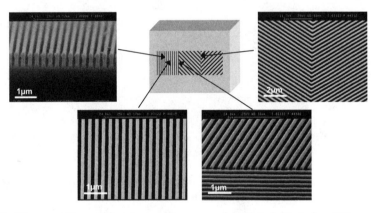

Fig. 8. The u-profile metal stripe grating array fabricated. Period = 300 nm, metal = chromium, substrate = fused silica.

In our case, we used e-beam writing for the generation of matrix arranged fields of different grating directions and reactive-ion-beam etching for its transfer 600 nm into the quartz substrate. After this, 30-nm chromium was coated with substrate rotation under a deposition angle of 45°. With such a u-profile, metal stripe grating polarization ratios >300 (max. 1500 at 4525 nm) were measured in the range from 425 nm to 1000 nm wavelength. Gratings fabricated can be seen in Figure 8.

Application and fabrication methods of the binary subwavelength pattern are also described in Refs. [1–9].

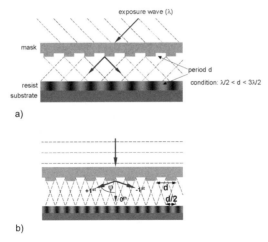

Fig. 9. Scheme of near-field holography with (a) oblique incidence illumination and (b) normal incidence illumination.

3. It is clear that many binary optical elements are based on gratings with fine pitch. Holography or electron-beam writing can be used for its generation of course. Both methods have shown its potential for high resolution at good quality. Nevertheless, e-beam writing is often criticized because of its stitching errors between work fields and because of the low writing speed that results in high fabrication costs. Both problems have been reduced in the past by special writing strategies for instance. A total solution of these problems cannot be expected in the near future. The handicap of free-space holography is its limited flexibility for arbitrary grating fabrication. Especially developed for fine-pitch grating replication is near-field holography (NFH). This lithographic method sketched in Figure 9a tries to exploit the advantages of e-beam writing and holography. Behind the oblique illuminated mask that is fabricated by e-beam writing, for instance, the zeroth and the first diffraction order interfere. The period of interference is the same as the grating period of the mask and does not depend on the illumination angle or the illumination wavelength. Under a fixed illumination wavelength (e.g., 365 nm), boundary conditions have to be fulfilled and optimizations have to be done for a successful or optimized operation of this kind of lithography:

- Two beam interference ($\rightarrow$ periods that can be copied with NFH)
- Symmetric angles of the diffraction orders ($\rightarrow$ illumination angle)
- Equal intensities of the diffraction orders ($\rightarrow$ mask design)

Amplitude masks and phase masks can be used as well for this version of NFH. By softening the second and partially the third condition, it is possible to copy a period range instead of a single period. To overcome the restriction regarding the grating direction, normal incidence of illumina-

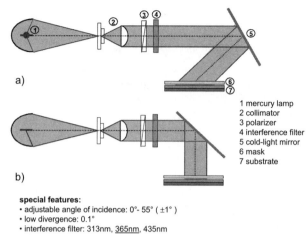

a)

b)

1 mercury lamp
2 collimator
3 polarizer
4 interference filter
5 cold-light mirror
6 mask
7 substrate

special features:
• adjustable angle of incidence: 0°- 55° (±1°)
• low divergence: 0.1°
• interference filter: 313nm, 365nm, 435nm

Fig. 10. Scheme of the mask aligner MA6 NFH for near-field holography. (a) for normal and (b) for oblique incidence.

tion can be used (Fig. 9b). In this case, unpolarized or circular polarized and a careful designed phase mask that suppresses the zeroth diffraction order is necessary. This leads to interference fringes with half the period of the original mask pattern. The design of the masks has to consider polarization problems also. If unpolarized light is used for copying the mask, diffraction orders of both the TE and the TM polarizations should fulfill the specs as well. Because the phase mask must be feasible, this turns out to be difficult sometimes. NFH is a typical lithographic method that was developed preferably for optical pattern fabrication – in this case fine pitch grating. The original idea of its application was the fabrication of gratings for distributed feedback (DFB) lasers. Figure 10 shows the scheme of the commercial mask aligner MA6 NFH from Suss Microtec. Now, a further stage of NFH is able to copy or generate many more types of gratings. Figure 11 shows a couple of examples [10].

3.2 Fabrication Methods for Multilevel Patterns

Optical element design often ends in multilevel phase elements. The reason for this might be the wish for an approximation of refractive elements by stairs, the need of an maximized diffraction efficiency of a diffractive element, the increase of the uniformity of an beam splitter, or others. The elements that result are multilevel heath profiles for transmission or reflection.

One of the most common fabrication methods for such elements uses a multiple overlay of the well-established technique for binary pattern fabrication. A doubling of the height levels can be done with each binary step. A scheme of this method is seen in Figure 12. It is easy to recognize that the number of height levels increases very fast with the binary fabrication steps: N

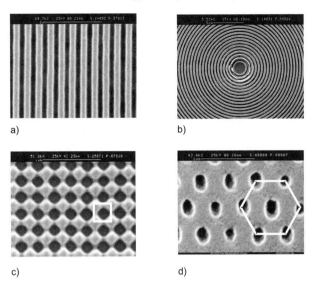

Fig. 11. Examples of grating fabricated by near-field holography: (a) 400-nm period grating; (b) circular grating 500 nm period; (c) 500-nm period grating made by two exposure steps crossed under 90° and (d) under 60°.

n masks ➜ 2^n levels

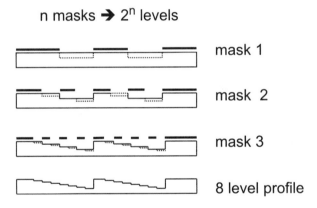

Fig. 12. Scheme of the fabrication of multilevel optical elements.

steps generate 2^N levels. Photolithography, electronlithography, or other lithographic methods are suitable for the etching anisotropic; dry etching should be preferred. Figure 13 shows details of a diffractive element fabricated by e-beam lithography and reactive-ion-beam etching in fused silica. The element of Figure 13a is based on a 400-nm × 400-nm pixel size and includes four levels Figure 13b shows a phase dislocation element with eight levels.

Because advantages are expected theoretically by the use of a high number of phase levels, the fabrication is pushed to make a large number of height levels. Due to the limited overlay accuracy and the nonideal side walls of the

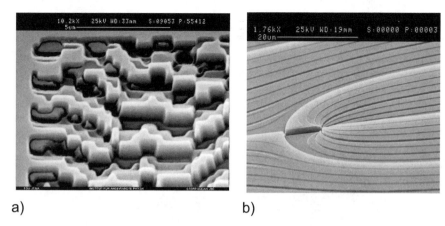

a)   b)

Fig. 13. Multilevel diffractive elements fabricated by e-beam direct writing and reactive-ion-beam etching in fused silica: (a) nonparaxial beam-splitting element; (b) phase dislocation.

etched pattern, in reality an optimum of levels exists. On the one hand, this can be understood as the sidewalls and the overlay error of multiple lithographic steps that leads to a disturbed region and may occupy a substantial part of the element area. That area decreases the diffraction efficiency or the optical signal quality. This means that a large increasing of the number of height levels might lead to the destruction of the optical element as well as its optical function. On the other hand, the optimum cannot be described as easily by the proportion of the destructed element area. This depends strongly on the type of optical element. For instance, the uniformity of the diffraction order intensity of beam splitters is very sensitive to overlay errors, even at small deflection angles. Less sensitive are Fresnel lenses in the paraxial domain. The patterns of optical elements disturbed by overlay and sidewall problems are shown in Figure 14. The pattern shown in Figure 14a should look similar to Figure 13a. Sidewall and overlay problems destroyed the pattern. As the lithography for the phase dislocation element of Figure 14b was done with a mask aligner that reaches just 1 μm overlay accuracy, undesired walls have been engendered within the element fabrication. Figure 15 helps in the understanding of the connection between the overlay error and the walls or gaps. Usual values for overlay errors are 0.5 to 1 μm for photolithographic mask aligners, 50 to 200 nm for e-beam writer or optical wafer stepper, and below 50 nm for advanced lithography tools (wafer stepper or e-beam writer). A good estimation of the loss in diffraction efficiency due the fabrication errors is given in the work of Ricks [11].

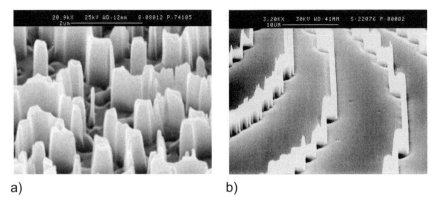

a)   b)

Fig. 14. Pattern of multilevel diffractive elements disturbed by overlay errors: (a) DOE similar to Figure 13a; (b) detail of a phase dislocation pattern similar to Figure 13b.

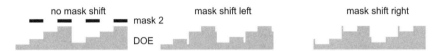

Fig. 15. Gaps and walls can be engendered by overlay errors. Sketched is the second mask application and the consequence of its position shift to the left and to the right.

3.3 Fabrication Methods for Continuous Profiles

Resist Melting/Reflow

The resist melting technique is a very well established and simple method for the fabrication of spherical or cylindrical lenses (see [12–16]). Its principle is shown in Figure 16. At first, cylinders of photoresist with the appropriate base area and volume have to be generated on the substrate by binary photolithography. Afterward, the resist cylinders will be heated up on a hot plate. Above its glass transition temperature T_G, the photoresist changes into a viscous state and surface tension leads to the formation of a surface of least energy (surface of minimal area). As an example, the temperature T_G for the common diazonaphtoquinon (DNQ) resists is in the range 70°C to 110°C (dependent on the molar mass).

For the ideal melting process without constraint forces, the minimal surface is defined by the resist volume only. The rim angle in this case is given by the interfacial tensions σ_{ij} of substrate material (index 1), resist (index 2), and atmosphere (index 3):

$$\cos(\alpha) = \frac{\sigma_{13} - \sigma_{12}}{\sigma_{23}}. \tag{1}$$

In practice, the base area of the photoresist cylinders is fixed and the rim angle can even be larger than the characteristic value. Therefore, both parameters

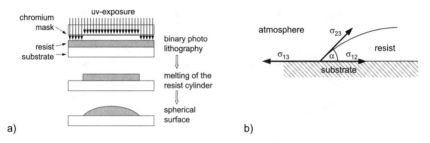

Fig. 16. (a) Principle of the resist melting process and (b) the interfacial tensions at the rim of the resist, defining the characteristic rim angle α.

(resist volume and base area) define the minimal surface. For example, photoresist cylinders with a circular base area are shaped into spherical lenses. The radius of curvature R of such a lens can be calculated from the following equations. The volumina V_{lens} and V_{cylinder} are assumed to be proportional:

$$V_{\text{lens}} = \frac{\pi}{6} h \left(3r^2 + h^2 \right) = f \pi r^2 d = f V_{\text{cylinder}},$$

with r and h being the radius and the height of the resist cylinder, respectively, and d being the height of the resist cylinder. The factor f takes into account that the resist volume decreases during the melting process due to cross-linking of the photoresist and the evaporation of the residual solvent. The relation among r, h, and R is

$$R = \frac{r^2 + h^2}{2h}.$$

The rim angle of the melted resist lens can vary in a wide range (also ball lenses are possible), but the characteristic value given by Eq. (1) was found to be the lower limit. If the volume is too small, the fabricated profile is not a spherical one, but has a flat top or even a dip in the central part. As a consequence, the numerical aperture NA of melted lenses has a lower limit of $\text{NA}_{\text{min}} \approx 0.1$ [14].

This drawback can be overcome with the reflow technique, which uses a solvent atmosphere at room temperature [17]. Under these conditions, the resist becomes liquid due to diffusion of the solvent into the polymer matrix. The characteristic rim angle in this case is very low ($< 1°$) and the resist will not stick to the substrate, as for the thermal melting. In order to avoid spreading of the resist across the substrate, pedestals with the shape of the desired base area has to be used. This can be done using binary photolithography and a subsequent transfer of the resist pedestals into the substrate material by dry etching (Fig. 17). In contrast to the thermal melting, the photoresist remains sensitive to UV exposure. This property makes the reflow process very attractive, because a second lithographic step can be added directly onto the minimal surface.

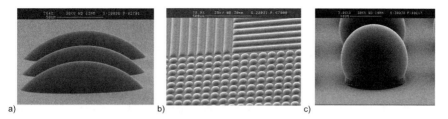

a)   b)   c)

Fig. 17. Examples of spherical (and cylindrical) lenses fabricated with (a) the resist melting technology, (b) the reflow technique, and (c) the glass melting method.

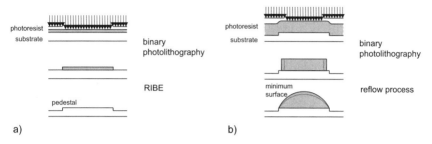

a)   b)

Fig. 18. Sequence of the reflow technique. After etching of the pedestals (a) photoresist will be shaped to a minimal surface in a solvent atmosphere (b).

It should be mentioned that there are also other techniques for the fabrication of microoptical elements (microlenses) using the surface tension. The ink-jet lens printing for instance utilizes the ink-jet printer technology ("drop on demand") for the production of small lenses with very good reproducibility [18, 19]. Instead of ink, a well-defined amount of a UV-curing material (e.g., hybrid organic–inorganic sols with photoinitiator) will be placed onto the substrate. Since the drops are rotationally symmetrical, the emerging minimal surfaces will be spherical lenses of certain size and volume. After the polymerization by UV curing, the material is optically transparent and mechanically stable. With an appropriate positioning system, the single lenses can be placed to form arrays of arbitrary pattern. Also, the generation of larger elements is possible by placing a number of lenses close together and letting them merge.

A promising method for the direct fabrication of glass lenses on a fused silica substrate uses a special glass (Boron-Phosphorus-Silicate Glass, BPSG) of low melting temperature instead of the photoresist [20]. The BPSG will be deposited onto the substrate with a special flame hydrolysis process, and with binary photolithography and dry etching, cylinders of BPSG are generated. At a temperature between the melting points of BPSG and fused silica, the cylinders will melt and form the minimal surface.

Examples of lenses fabricated with the different technologies are shown in Figure 18.

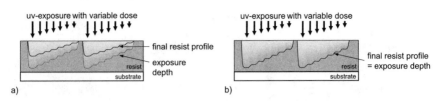

Fig. 19. Control of the final resist profile in analog photolithography. In the conventional process, the development is stopped after a certain time (a). The proposed alternative method of development does not until the resist profile has reached the exposure depth (b).

The advantages of the resist melting or reflow technique are its simplicity and the good optical performance of the fabricated lenses due to the good sphericity and very smooth surface. However, because these methods use surface tension for the profile generation, the variety of microoptical elements which can be fabricated is very limited. There exist no other minimal surface profiles than spherical and cylindrical. Even if the base area is chosen so that its rim is a line of constant height of the desired surface profile, the generated profile cannot be used directly as an optical element. Therefore, methods for the fabrication of almost arbitrary surface profiles are of great importance and we want to focus on it in Section 3.4.

Analog Lithography

With the term "analog lithography," all of the lithographic methods should be summarized which make use of the analog dependency of the dissolution rate of the resist on the exposure dose. This can be either the UV exposure of a photoresist or the electron-beam exposure of an electron resist. During the development process, the locally varying dissolution rate leads to the formation of a profile and the development is stopped if the desired surface profile is reached (see Fig. 19a). Thus, the final profile depth is controlled by the exposure dose and the development time.

For photolithography there is also an alternative approach of development which makes use of the absorption of the photoresist [21]. Since the intensity of the exposure wave decreases exponentially within the resist layer, the photoresist cannot be exposed completely from the surface to the bottom if the resist thickness exceeds a certain value. In this case, a residual layer of resist remains unexposed and the boundary between exposed and unexposed regions of the resist (exposure depth) is dependent on the exposure time. In contrast to the conventional process, the development can be continued until the profile has reached the exposure depth (see Fig. 19b). In this case, the final profile is dependent on the exposure dose only. Moreover, the remaining resist is unexposed and this gives the possibility of applying a second lithographic step directly onto the resist profile.

The different methods of analog lithography can be distinguished by the following working principle: First, writing technologies (*laser beam writing* and *e-beam writing*) use a small beam which will be scanned across the substrate in order to expose the pattern sequentially. Second, printing technologies (for UV exposure) use a mask to modulate the uniform intensity distribution of the illumination wave and the desired pattern will be exposed in parallel. In this case, two different mask types resulting in *halftone lithography* and *gray-tone lithography* can be distinguished.

Halftone Lithography

The basic idea of halftone lithography is to use a binary mask with high-resolution structures which generate the different gray levels. In principle, this task can be realized by two-dimensional amplitude gratings (Fig. 20). For such gratings, the intensity of the zeroth transmitted diffraction order is given in first-order approach (no resonance effects) by the fill factor of the grating. The fill factor is defined as the ratio of the covered area to the whole area of one grating period. There are different methods of controlling this fill factor, as the covered area can be varied and the period is fixed (pulse-width modulation) or vice versa (pulse-density modulation). Also, the random distribution of covered area and period is possible (error diffusion). In any case, it is necessary to provide a spatial filtering of the illumination wave behind the halftone mask. Since the period is larger than the wavelength for exposure, the halftone structures will act as gratings with more than one propagating diffraction orders. Because of the interference of these diffraction orders, the intensity distribution behind the mask will be totally different from the desired one. As a consequence, contact printing of halftone masks in a mask aligner does not provide the desired results. However, if the mask plane is imaged into the wafer plane, as, for instance, in a wafer stepper, the optical imaging system provides a spatial filtering. The cutoff frequency of this filter is the reciprocal of the minimal resolvable feature size d_{min}, which can be calculated from Abbe's theory of diffraction:

$$d_{\mathrm{min}} = k \frac{\lambda}{\mathrm{NA}}.$$

Here, λ is the exposure wavelength and NA is the numerical aperture of the objective lens. The factor k, which ranges between 0.5 and 1.0, is dependent on the kind of illumination wave (e.g., angle of incidence, degree of coherence). If the grating period is chosen so that it will not be resolved by the imaging system, only the zeroth diffraction order will be transmitted, carrying the information of the fill factor.

Halftone lithography and its application is described, for instance, in Refs. [22] and [23] and Figure 21a shows an example of a fabricated surface profile. Since the halftone masks will be fabricated by electron-beam lithography preferably, the accuracy of halftone lithography is related to the accuracy

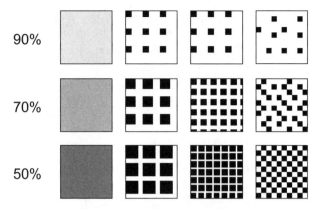

Fig. 20. Different gray levels (left column) and its representation by different coding principles: pulse-width modulation, pulse-density modulation, and error diffusion (from second left to right column).

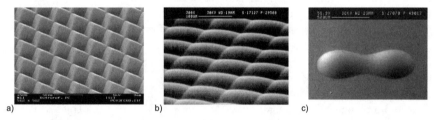

a)   b)   c)

Fig. 21. Examples of different microoptical elements fabricated with (a) halftone and (b, c) graytone lithography: Array of retroreflectors (courtesy of K. Reimer, Fraunhofer ISIT, Itzehoe), array of convex lenses (size of each lenslet: 100 μm × 100 μm, height: 10 μm) and a beam-shaping element (size: 1200 μm × 600 μm, height 42 μm).

of the e-beam machine. The influence of the absolute placement and pattern error on the accuracy of the fill factor increases with decreasing period. If, for instance, a grating feature is inaccurate by 50 nm within the period of 500 nm, this leads to an error of the transmission value of 10%.

Gray-Tone Lithography

The gray-tone lithography uses a special mask that locally controls the transmission of the illuminating light in photolithography. Commercially available is the so called HEBS glass (High-Energy-Beam-Sensitive glass) [24–26]. Essentially, this is an ion-exchanged white crown glass with various additives. It contains silver-alkali-halide complex crystals, which have a size of about 10 nm or less in each dimension within a sensitive layer of 3–10 μm thickness. Chemical reduction of the silver ions to clusters of atomic silver can be induced, for instance, by electron-beam exposure. Since the size and the distance of the clusters are far below the exposure wavelength of lithography, they only induce a change of the complex index of refraction, but no stray light or propagating diffraction orders will appear. Caused by the presence of silver atoms, the absorption coefficient of the glass increases in a wide spectral range, which is observed as a darkening. Additionally, the glass is doped with special photoinhibitors, which increases the energy band gap of the otherwise photosensitive glass and allows the use of the darkened glass as a photomask for the photolithography ($\lambda > 350\,\mathrm{nm}$). The achievable transmission values (relative to the unexposed glass) range from less than 0.01 to 1.0 at 365 nm wavelength.

The principle of the gray-tone lithography is shown in Figure 22. At first, the gray-tone mask will be generated by an electron-beam exposure with a variable dose. The locally implemented electron dose controls the optical density of the mask, which can then be applied in photolithography. The big advantage over halftone masks is the possibility of using the HEBS glass mask in a mask aligner, which is a quite inexpensive tool. In the second step, the photoresist will be exposed by the illumination wave, the intensity of which was modulated by the HEBS glass mask. In dependency on the UV exposure dose, the dissolution rate of the resist will be controlled and finally, after development, the surface profile of the developed resist. To sum up, the final resist profile is controlled by the electron dose which was exposed to the HEBS glass mask. This dependency is shown in Figure 23 for different initial resist thicknesses of the AZ4562 photo resist (Clariant GmbH). Each calibration curve corresponds to a certain set of processing parameters, such as resist baking time and temperature, exposure time, and development time. Any change in the parameters influences the calibration and it needs to be measured prior to the fabrication of a desired surface profile. The maximum profile depth which can be achieved with a certain photoresist is limited by its viscosity, absorption, and sensitivity. As can be seen from Figure 23, with the AZ4562 resist a maximum profile depth of 65 µm can be reached. With other materials, even deeper profiles are possible [27]. Some examples of fabricated refractive elements are shown in Figures 21b and 21c.

As discussed in Section 2, the accuracy of the generated surface profile is essential for the fabrication of refractive microoptical elements, but it is important for diffractive elements with a continuous profile too. Investigations of the HEBS glass have shown different effects which influences the accuracy of the fabricated profiles. Because of the dynamics of e-beam exposure of the HEBS glass, the local gray-tone is dependent not only on the electron dose but also on the method of exposing this dose [28]. This induces systematic deviations of the generated gray-tone distribution to the desired one in dependency on the exposed pattern. The final resist profile, therefore, will show a mask-induced roughness in addition to the intrinsic roughness of the photoresist. Another effect which has to be considered is the correlation between the induced absorption of the HEBS glass and the change of the refractive index. For a transmission value of 0.01, the index change is about 10^{-2}. Thus,

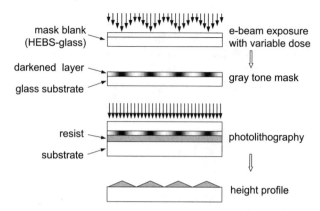

Fig. 22. Principle of the gray-tone lithography with HEBS glass.

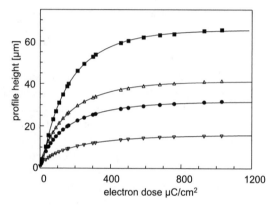

Fig. 23. Dependency of the final height of the resist profile on the electron dose of e-beam exposure of the HEBS glass mask (calibration). Each curve represents the calibration for a certain set of process parameters.

the gray-tone mask simultaneously modulates the phase of the illumination wave.

However, the gray-tone lithography has developed to an established technology for the fabrication of continuous surface profiles. The accuracy of fabricated elements is below 1% of the total profile depth, and for the array of spherical lenses in Figure 21b, the total deviation of the profile to the desired shape was only 0.3%.

Laser Writing and e-Beam Writing

As mentioned earlier, analog lithography also includes writing technologies: laser writing [29–32] and e-beam writing [33, 34]. The equipment for both technologies is quite similar, as sketched in Section 3.1 and also the principal method of exposure is the same. The beam can be focused into a tight spot at the resist surface or it can be the reduced image of an aperture. In the latter case, geometrical primitives (circle, rectangle, triangle) or even structural cells (e.g., part of a grating) can be exposed in parallel, which clearly decreases the exposure time. Furthermore, the beam has to be scanned across the substrate, and for this purpose, either the beam can be deflected or the x/y-stage with the substrate can be moved. Also, the combination of beam deflection in small working areas with table movement for larger distances has been realized. Control of the exposure dose can be achieved by dwell time or by layer addition (variable dose writing).

However, due to the different physical nature of the exposing radiation, laser and e-beam writing differ in some principal parameters. For the focused beam, the minimal spot size as well as the depth of focus depend on the wavelength of the laser beam and on the electron energy for the e-beam, respectively. Since the laser beam has a spot size of about 100 to 500 nm, the electron beam may be focused to a spot 1 nm in size. The depth of focus depends on the NA (about 0.5 for laser writing and about 0.001 for e-beam writing) (Fig. 24). For the laser beam, it is a depth of about 1 μm, and for the e-beam, it is usually tens of micrometers. However, the electrons have a limited reach of several micrometers in the resist, while the light can penetrate the photoresist according to the absorption and the intensity. Exposure that causes bleaching of the photoresist can also increase the penetration depth of the light. Light-scattering effects in photoresists are negligible and do not reduce the resolution of laser writing, even in case of deep profile fabrication in thick resist layers. The scattering of electrons is an essential effect. This scattering leads to significant blurring of the electron beam as it propagates through the resist layer and thus to a loss of lateral resolution. Finally, laser-beam writing and e-beam writing present different features for analog lithography. Laser-beam writing can be used for fabricating large profile depth (> 100 μm) with a resolution behind the e-beam; e-beam writing can be used up to several micrometers profile depth (depends strongly on the electron energy) and shows very high resolution for shallow profiles (< 1 μm) only.

3.4 Fabrication Methods for Multifunctional Surface Profiles

Fabrication of multifunctional surfaces means the fabrication of a surface that can be divided into at least two functionalities descending from different classes (or subclasses) of optical pattern to be fabricated with different fabrication methods. Such a strategy is known in principle from the hologram fabrication techniques, for which different exposures are added sequentially for different optical functions. For surface profile elements the challenge is to superimpose several structures that vary significantly in depth and lateral resolution. In the following, we give some ideas regarding the fabrication of such elements.

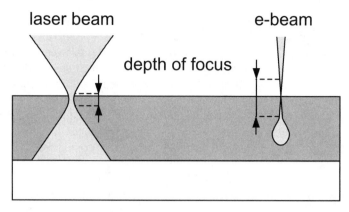

Fig. 24. Schematic comparison of the numerical aperture and depth of focus of a laser beam and an electron beam. The light penetrates the resist, whereas the electron beam has a limited reach.

Pattern on Multilevel Profiles

One version for combining different profiles is the use of a prestructured substrate for the multilevel element fabrication process described earlier. Such a strategy makes sense, for instance, if moth eye patterns for antireflection purposes are desired on a binary or multilevel pattern. Figure 25 shows this strategy and preliminary results. The moth eye pattern can be fabricated by e-beam writing or holographic techniques (e.g., NFH) and reactive-ion-beam etching into fused silica. After this follows the fabrication of the next coarser binary level that contributes to the final profile. If a four-height-level pattern has to be fabricated, it follows the last and coarsest lithographic level. Each etching step applied transfers the uncovered part of the previous profile fabricated into the substrate. Ideal would be an etching without any deformation of the uncovered profile. Therefore, this etching is an subject for optimization in any case [10].

Pattern on Continuous Profiles

As mentioned earlier with respect to the reflow technique or analog photolithography, it is possible to fabricate continuous surface profiles with a large depth and high accuracy. The reflow process is used to fabricate a minimal surface which has only small deviations to the desired surface profile. However, this property of the reflow process limits the freedom to fabricate arbitrarily shaped profiles. Much more freedom regarding the profile shape gives the gray-tone lithography. Nevertheless, high-resolution patterns on large depth continuous surfaces are impossible. The way out is to use the profiles fabricated by reflow or gray-tone lithography as a preform for further lithographic steps with a high resolution that is sufficient for the pattern to add [34–37].

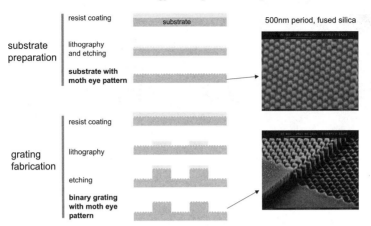

Fig. 25. Principle of the fabrication of a binary (multilevel) profile with a subwavelength structure on top. The second step of the binary profile fabrication can be repeated to generate a multilevel profile.

This can be done on reflow fabricated preforms without problems. In contrast to the thermal melting, the photoresist remains sensitive to UV exposure after the reflow process. This property makes the reflow process very attractive. Preforms fabricated by analog lithography with exposure depth control can be used for further lithographic steps as well. Exposure depth control is a version of analog photolithography described earlier. It uses the absorption in combination with the bleaching of the resist for controlling the exposure depth. With the development process, all of the exposed resist will be removed and unexposed resist is the profile wanted.

An alternative method to make preforms sensitive for further lithographic steps, especially for e-beam writing, is an overcoating. Spin coating with optimized parameters can be used, for instance. This might be helpful because resists for photolithography are not as good for e-beam lithography.

Figure 26 shows examples of multifunctional surfaces fabricated. E-beam lithography features with high resolution and high depth of sharpness but laser writing can be used for less resolution pattern, too.

Profile Correction

On reflow preforms, analog photolithography can be applied directly onto the minimal surface in order to correct for the deviations. The advantage of this preform technique is an increased accuracy of the final element in comparison to the same element fabricated with analog photolithography only. This is due to the fact that the minimal surface is defined by the physical effect of surface tension. With the given base area and resist volume, it can be calculated very precisely. Because, in the second step, the analog photolithography has to

a)   b)   c)

Fig. 26. Examples of multifunctional surface profiles: (a) refractive microlenses with a fresnel lens, (b) a linear grating, and (c) a diffractive beam shaping element on top. Pictures (b) and (c) courtesy of U. Zeitner, Fraunhofer IOF, Jena.

realize a small profile depth, the profile error is small also. Thus, the total profile error is reduced to the error of the correction profile.

For example, a beam-shaping element could be fabricated with the preform technique. In a first step, a reflow profile has to be fabricated that is as near as possible to the profile wanted. On this occasion, the variation freedom of reflow profiles should be exploited maximally. This includes preforms resulting from ellipse- or kidney-shaped footprints too [35, 38].

4 Conclusions

The goal of this chapter was to present the large variety of microstructure technologies for optical component fabrication. Nevertheless, only a small part of fabrication techniques and technologies for optics has been realized. Up to now, many elements have been fabricated for the practical demonstration of optical effects predicted theoretically or has been designed on the base of engineering. Also, industrial applications of microstructured optical elements exist for a long time and become more and more important. However, the breakthrough of this kind of optics is strongly connected with the quality and the costs of the elements, with compatibilities and with its system ability. With regard to this, there is certainly plenty of room for further activities and improvement. Especially nanooptics, not considered here but in Chapter 10 by Jamois et al., will challenge the technology regarding feasibility and economic fabrication in the future.

Acknowledgments

The authors would like to thank all of the colleagues of the microoptics/micro-lithography group of the Institute of Applied Physics of the Friedrich-Schiller-University, Jena for their long-term collaboration and contribution to this chapter. We also would like to thank the colleagues outside the institute who contributed with figures for illustrating the state of the art in microoptics technology.

References

1. Kley, E.-B., Fuchs, H.-J., and Zoellner, K., Fabrication technique for high aspect ratio gratings, Micromachining and microfabrication , Proc. SPIE 3879, 71–78 (1999).
2. Clausnitzer, T., Fuchs, H.-J., Kley, E.-B., Tuennermann, A., and Zeitner, U. D., Polarizing metal stripe gratings for micro-optical polarimeter, Proc. SPIE 5183, 8–15 (2003).
3. Gombert, A., Glaubitt, W., Rose, K., Dreibholz, J., Blasi, B., Heinzel, A., Sporn, D., Doll, W., and Wittwer, V., Subwavelength-structured antireflective surfaces on glass, Thin Solid Films 351, S.73–78 (1999).
4. Gombert, A., Glaubitt, W., Rose, K., Dreibholz, J., Blasi, B., Heinzel, A., Sporn, D., Doll, W., and Wittwer, V., Antireflective transparent covers for solar devices, Solar Energy 68(4) 357–360(Pt. A) (2000).
5. Haidner, H., Kipfer, P., Stork, W., and Streibl, N., Zero-order gratings used as an artificial distributed index medium, Optik 89, 107–112 (1992).
6. Lalanne, Ph., Astilean, S., Chavel, P., Cambril, E., and Launois, H., Blazed-binary subwavelength gratings with efficiencies larger than those of conventional Echelette gratings, Opt. Lett. 23, 1081–1083 (1998).
7. Lalanne, Ph., Astilean, S., Chavel, P., Cambril, E., and Launois, H., Design and fabrication of blazed-binary diffractive elements with sampling periods smaller than the structural cutoff, J. Opt. Soc. Am. A 16, 1143–1156 (1999).
8. Raguin, D. H., and Morris, G. M., Antireflection structured surfaces for the infrared spectral region, Appl. Opt. 32(7), 1154–1167 (1993).
9. Schnabel, B., and Kley, E.-B., Fabrication and application of subwavelength gratings, Proc. SPIE 3008, 233–241 (1997).
10. Kley, E.-B., and Clausnitzer, T., E-beam lithography and optical near field lithography: new prospects in fabrication of various grating structures, Proc. SPIE 5184, 115–125 (2003).
11. Ricks, D. W., Scattering from diffractive optics, in *Diffractive and Miniaturized Optics*, edited by Lee, S. H., Critical Reviews of Optical Science and Technology, SPIE Optical Engineering Press, Bellingham, WA, 87–211 (1993).
12. Daly, D., Stevens, R. F., Hutley, M. C., and Davies, N., The manufacture of microlenses by melting photoresist, J. Meas. Sci. Technol. 1 (Suppl.) 759–766 (1990).
13. Jay, T. R., Stern, M. B., and Knowlden, R. E., Effect of microlens array fabrication parameters on optical quality, Proc. SPIE 1751, 236–245 (1992).
14. Nussbaum, Ph., Volkel, R., Herzig, H. P., and Dandliker, R., Microoptics for sensor applications, Proc. SPIE 2783, 1081–1083 (1995).
15. Popovic, Z. D., Sprague, R. A., and Connell, G. A. N., Technique for monolithic fabrication of microlens arrays, Appl. 27(7), S.1281–1284 (1988).
16. Haselbeck, S., Schreiber, H., Schwider, J., and Streibl, N., Microlenses fabricated by melting photoresist on a base layer, Opt. Eng. 32(6), 1322–1324 (1993).
17. Erdmann, L. and Efferenn, D., Technique for monolithic fabrication of silicon microlenses with selectable rim angles, Opt. Eng. 36(4), 1094–1098 (1997).
18. Biehl, S., Danzebrink, R., Oliveira, P., and Aegerter, M.A., Refractive microlens fabrication by ink-jet process, J. Sol-Gel Sci. Technol. 13, 177–182 (1998).
19. Cox, W. R., Chen, T., and Hayes, D.J., Micro-optics fabrication by ink-jet printing, Opt. Photonics News 12 (6), 32–35 June 2001.

20. Kley, E.-B., Fuchs, H.-J., and Kilian, A., Fabrication of glass lenses by melting technology, Proc. SPIE 4440, 85–92 (2001).
21. Wittig, L.-C., Clausnitzer, T., Kley, E.-B., and Tuennermann, A., Alternative method of gray-tone lithography with potential for the fabrication of combined continuous 3D surface profiles and subwavelength structures, Proc. SPIE 5183, 109–115 (2003).
22. Oppliger, Y., Sixt, P., Stauffer, J. M., Mayor, J. M., Regnault, P., and Voirin, G., One-step 3D shaping using a gray-tone mask for optical and microelectronic applications, Microelectron. Eng. 23, 449–454 (1994).
23. Reimer, K., Engelke, R., Hofmann, U., Merz, P., Kohlmann-von Platen, K. T., and Wagner, B., Progress in gray tone lithography and replication techniques for different materials, Proc. SPIE 3879, 98–105.
24. Daeschner, W., Long, P., Larsson, M., and Lee, S., Fabrication of diffractive optical elements using a single optical exposure with a grey level mask, J. Vac. Sci. Technol. B 13(6), 2729–2731 (1995).
25. HEBS glass photomask blanks, Product information 96-01, Canyon Materials Incorporation, San Diego, CA.
26. Wu, C. Methods of making high energy beam sensitive glasses, U.S. Patent 5, 078,771 (7 January 1992).
27. Rogers, J. D., Lee, J., Karkkainen, A. H. O., Tkaczyk, T., and Descour, M. R., Gray-scale lithographic fabrication of optomechanical features using hybrid sol-gel glass for precision assembly of miniature imaging systems, Proc. SPIE 5177.
28. Wittig, L.-Chr., Cumme, M., Harzendorf, T., and Kley, E.-B., Intermittence effect in electron beam writing, Poster auf Micro- and Nano-Engineering 2000 18.-21.09.2000, Microelectron. Eng. 57–58, 321–326 (2001).
29. Gale, MT. and Knop, K., The fabrication of fine lens arrays by laser beam writing, Proc. SPIE 398, 347–353 (1983).
30. Gale, M.T., Rossi, M., Pedersen, J., and Schutz, H., Fabrication of continuous-relief micro-optical elements by direct laser writing in photoresist, Opt. Eng. 33, 3556–3566 (1994).
31. Gale, M.T., Direct writing of continuous-relief micro-optics, in *Micro-optics: Elements, Systems and Applications*, edited by H.P. Herzig, Taylor & Francis, London (1997).
32. Goltsos, W. and Liu, S., Polar coordinate writer for binary optics fabrication, Proc. SPIE 1211, 137–147 (1990).
33. Fujita, T., Nishihara, H., and Koyama, J., Blazed gratings and Fresnel lenses fabricated by electron-beam lithography, Opt. Lett. 7, 578–580 (1982).
34. Kley, E.-B., Schnabel, B., and Zeitner, U. D., E-beam lithography: an efficient tool for the fabrication of diffractive and micro-optical elements, SPIE Proc. 3008, 222–232 (1997).
35. Wittig, L.-C., and Kley, E.-B., Approximation of refractive micro-optical profiles by minimal surfaces, Proc. SPIE 3879, 222–232 (1999).
36. Traut, S., and Herzig, H. P., Holographically recorded gratings on microlenses for a miniaturized spectrometer array, Opt. Eng. 39(1), 290–298 (2000).
37. Traut, S., Rossi, M., and Herzig, H. P., Replicated arrays of hybrid elements for application in a low-cost micro spectrometer array, J. Mod. Opt. 47(13), 2391–2397 (2000).
38. Kley, E.-B., Thoma, F., Zeitner, U.D., Wittig, L., and Aagedal, H., Fabrication of micro optical surface profiles by using gray scale masks, Proc. SPIE 3276, 254–262 (1997).

Microlithographic Pattern Generation for Optics

Bernd Schnabel

Leica Microsystems Lithography GmbH, Göschwitzer Str. 25, 07745 Jena,
Germany
bernd.schnabel@leica-microsystems.com

1 Introduction

For the fabrication of optical and microoptical elements like gratings, lenses,
waveguides, zoneplates, computer-generated holograms, and so on, many different fabrication technologies are available. Some of these elements are planar
elements; that is, the optical function of such elements is realized by a thin
planar layer containing special structures and/or materials. For the fabrication
of such elements, lithographic technologies are employed very often.

The interrelation between the function and quality of an optical element
and the fabrication technology applied is important for the further development of optics. In the current chapter, we focus on this cross-relation for
lithographic fabrication technologies. In a first part, an overview on several
lithography methods is given. Then, laser-beam writing and electron-beam
writing (as the two methods which are mainly used for commercial fabrication) are discussed in more detail. Finally, the current state of the art is
shown.

In the second part of the chapter, the advantages and drawbacks of lithography for the fabrication of optical elements are addressed. Since the development of lithography tools is actually strongly driven by the semiconductor
technology, the differences between the semiconductor domain and optics are
discussed from the lithography tool supplier's point of view.

2 Methods of Lithographic Pattern Generation

Lithography as a technology for fabricating planar structures with dimensions
ranging from millimeters to nanometers has developed in the last 30 to 40
years as a powerful technology. Today, its main commercial application is in
the semiconductor domain, and the techological development of lithography
was strongly driven by the demands of chip manufacturing. This will certainly
continue in the coming years.

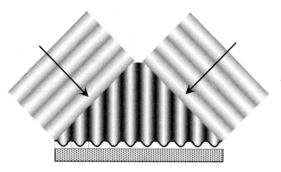

Fig. 1. Principle of grating exposure by interference of two plane waves.

In addition to semiconductor fabrication, lithographic methods are used in many other fields, like micromechanics, microoptics, or microbiology. However, these applications are mostly research-oriented and usually not as strong a business as semiconductor manufacturing. Nevertheless, the question of writing time is an important issue also for such applications because the fabrication costs are always of concern.

The question of exposure sequence is very important regarding both the productivity (writing time/throughput) and the flexibility of a lithographic technology. Several technologies exist where the entire pattern is exposed simultaneously (*parallel* writing techologies). This holds in particular for all mask-using technologies. As an advantage, complex patterns can be realized with a short exposure time. However, any change in the design of the pattern requires the fabrication of a new mask which is expensive and makes this technique quite inflexible.

A parallel but maskless technology, which is very important for optics, is the fabrication of large gratings by holographic intereference. Two interfering plane waves generate an interference pattern which is used for exposing the resist on a substrate (see Fig. 1). The gratings are formed either by the resulting surface profile (surface gratings) or by the change of the refractive index of the exposed material (volume gratings). As an advantage, with this method, gratings with very large areas (up to square meters) can be fabricated with a short exposure time. However, the realization of undisturbed plane waves with large extention (which are required to obtain error-free gratings) is extremely demanding, and the whole technology is limited to linear gratings.

As an alternative to parallel methods, *serial* lithographic technologies can be used. The sequential fabrication and the nonexistence of any mask allow a very high flexibility and the realization of arbitrary designs. Thus, such techniques (often referred to as "direct writing") are frequently used for optical patterns. The main problem of all sequential writing methods is the exposure time (i.e., the throughput), which essentially influences the fabrication costs. Thus, many approaches for througput enhancement exist.

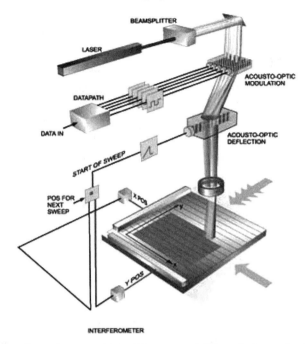

Fig. 2. Laser-beam writing. (Courtesy of Micronic Laser Systems AB.)

There are several direct-writing technologies using different kind of radiation for exposure, such as laser beams, X-rays, electron beams, and ion beams. However, the two mainly used methods nowadays are laser-beam writing and electron-beam writing, which will be briefly discussed. More detailed information on lithographic technologies can be found in Refs. [1] and [2].

2.1 Laser-Beam Writing

Laser-beam writing tools use optical radiation with wavelengths in the visible or ultraviolet (UV) domain for exposure. The resolution (minimum feature size) is essentially influenced by the wavelength and the numerical aperture (NA) of the optics, which results in an ongoing shortening of the wavelength and an increase of the NA. Actual tools use wavelengths of 248 nm or 193 nm. There are several suppliers of laser writing tools for volume manufacturing applications. As an example, the basic principle of Micronic's Omega6000 scanning lithography tool is shown in Fig. 2. Important here is the splitting of the laser beam into a bundle of independent beams, which enhances the throuput considerably.

Laser-beam writing was successfully applied for the fabrication of planar microoptical elements like gratings, fresnel lenses, and others, both using commercial systems or self-established setups [3, 4].

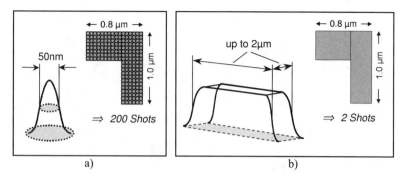

Fig. 3. Intensity distribution in the electron beam and exposure shot sequence for (a) Gaussian beam systems and (b) variable-shaped beam systems.

2.2 Electron-Beam Writing

Electron-beam (e-beam) lithography is established in semiconductor fabrication as the technology for high flexibility and highest resolution. There are many suppliers for professional e-beam tools in the world (but this chapter focuses on Leica electron-beam writing systems). Additionally, for some scanning electron microscopes (SEMs), add-ons are available, enabling low-cost and low-level e-beam lithography, especially for R&D applications. Usually electron energies between 50 keV and 100 keV are used, although low electron energies $E_0 \leq 10$ keV are sometimes employed for proximity-free lithography [5].

In general, there are two different approaches regarding the shape of the electron beam used for exposure: Gaussian beam systems and variable-shaped beam systems. Gaussian beam systems, which are basically similar to SEMs, use a circular beam with a Gaussian intensity distribution. The diameter of the focused electron beam is 2–5 nm for the best systems, enabling a very high resolution of the fabricated pattern. Feature sizes of less than 10 nm are possible [6, 7]. The problem of Gaussian beam systems is the writing speed, because for exposing larger areas, a very high number of subsequent shots is necessary (Fig. 3a).

For enhancing the throughput, electron-beam tools with a variable-shaped beam were developed. The electron beam is imaged into the target plane via a system of lenses and apertures. In the result, rectangular and triangular shots with variable size (up to several micrometers, depending on the system used) and orientation can be used for exposure (Fig. 4). In comparison to Gaussian beam systems, the number of shots is substantially reduced (see Fig. 3b).

A further measure for throughput enhancement is the so-called "write-on-the-fly" principle. Here, the exposure is performed while the stage and substrate are continuously moving. The exact alignment of substrate position and beam position is realized by the beam-tracking system. With this method,

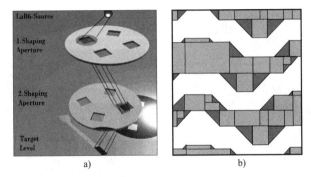

Fig. 4. (a) Principle of generating shaped electron beams; (b) using rectangular and triangular shots of arbitrary orientation for exposure of complex layouts.

a considerable shortening of the exposure time can be achieved because the time for settling the sample stage (which is required in conventional "stop-and-go" exposure after each sample stage motion step) is completely saved.

2.3 State of the Art

The increasing demands of semiconductor technology resulted in a rapid development of lithography in recent years. The improvement of parameters includes the following items:

- Resolution (minimum feature size)
- Substrate size
- Accuracy and stability parameters

The best resolution can be obtained with Gaussian beam e-beam tools due to the very small spot size. Lines with dimensions of about 10 nm wide can be fabricated (Fig. 5). With variable-shaped beam tools, features between 50 nm and 100 nm are realizable (see Fig. 6). For laser-beam writing tools, the minimum feature size is actually about 300 nm.

In parallel, the size of the substrates has enlarged continuously. For electron-beam writing tools, a limitation is given by the size of the vacuum chamber where the substrate and sample stage have to fit in. Nevertheless, e-beam writing on substrates up to 300 mm is available today (Fig. 6). Commercially available laser-beam writing tools allow for even larger substrates, up to 70 in. diagonal. As already mentioned in Section 2, the holographic exposure is capable for very large areas up to square meters.

The accuracy and stability of the lithography tools has improved as well. Most demanding application nowadays is the high-level mask writing, where writing times of 10 h and more for one single mask (130 nm node or 110 nm node) are typical. The question of accuracy will be discussed in more detail in Section 3.2.

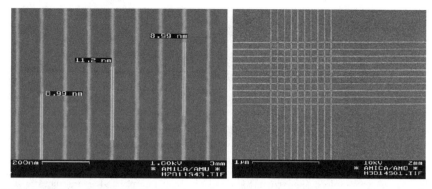

Fig. 5. Thin metal lines (width about 10 nm) fabricated by electron-beam direct writing (Leica VB6 system).

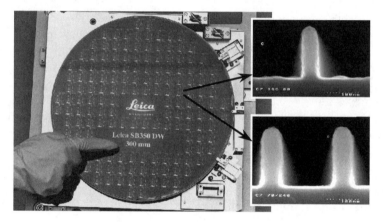

Fig. 6. High-resolution shaped e-beam writing on 300-mm wafers (Leica SB350DW system, top right: 60 nm isolated line; bottom right: 70 nm/170 nm lines and spaces. Exposure courtesy of LETI–Grenoble.)

3 Lithography and Optics

The preceding sections have shown that microlithography nowadays is a powerful tool. It is widely used for fabricating optical elements ([8–11] and chapter 2) (Figure 7 shows one example) and is often the technology of choice, due to its advantages (flexibility, resolution, and accuracy). On the other hand, the use of lithography tools for microoptics also shows some obstacles and challenges. This is mainly due to the fact that lithography tools are more and more specially adapted to the demands of semiconductor fabrication. The questions of how the development of lithography for optics will proceed and which problems and tasks have to be solved along the way arise. The following sections will address this.

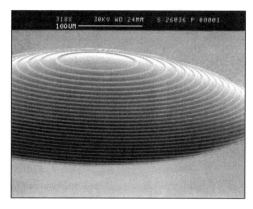

Fig. 7. Hybrid refractive–diffractive lens fabricated by electron-beam direct writing on a spherical profile obtained by melting. (Courtesy of Friedrich Schiller University, Jena.)

3.1 The Semiconductor Road Map

Although lithography is used in many fields besides semiconductor fabrication (like optics, micromechanics, biology, and others), the actual development of lithographic technology is strongly related to the semiconductor domain. For this, there are two main reasons:

1. The semiconductor fabrication is a strong business and most commercial lithography tools are sold to customers in this area, whereas optical applications of lithography often still remain in the R&D state.
2. For further technological development in the semiconductor domain, there is a clearly defined road map from where demands on and parameters for lithography tools can be unambiguously derived for the next years. In optics, there is no such road map.

Compared to recent forecasts, the semiconductor road map has actually accelerated and became more aggressive (Fig. 8). For commercial suppliers of lithography tools this is a very challenging evolution. As a result, the development of lithographic technology in the future will be more and more adapted to the special demands of semiconductor fabrication. The versatility of lithography tools will partially vanish, and the price for such sophisticated tools will certainly rise considerably.

3.2 Optics Versus Semiconductor Technology

As we have seen in the last section, optical applications have actually no major influence on the development of lithographic technology. Thus, there is the question of whether optics at least may *benefit* from the semiconductor-driven development of lithography.

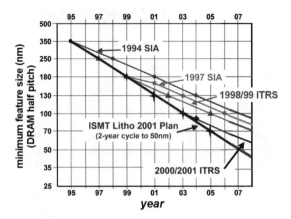

Fig. 8. Acceleration of technology road map in semiconductor fabrication.

Of course, the improvements of lithography tools are, in general, advantageous for optics. For example, the enhanced resolution allows the fabrication of optical subwavelength features, and accuracy improvements result in optical elements with better performance and fewer errors.

However, there are also many potential obstacles. Most of these issues arise from different requirements of semiconductor fabrication and optics. This include the following:

- Substrate size
- Substrate thickness
- Shape and orientation of the pattern
- Input data and data conversion for exposure
- Profiles
- Relation between function of the fabricated pattern and fabrication costs
- Error budget
- Characterization of the lithography tool performance

The versatility loss which accompanies the increasing adaption of lithography tools to the special demands of semiconductor fabrication will certainly increase interest in such issues in the future. Some of the points mentioned above will be discussed in more detail now.

Substrate Size and Thickness

For semiconductor applications, the size and thickness of the substrates are higly standardized. This is due to the fact that for cost-effective mass fabrication, standardization is a "must" and the substrates were adapted to meet the demands of fabrication technology. In mask writing, the most common mask size is 6 in. but 4, 5, or 7 in. is used too. The wafer size for direct-writing applications ranges from 100 mm, or 150 mm, for GaAs wafers to 150 mm,

200 mm, or 300 mm for Si wafers. The thickness values are fixed according to the substrate size as well.

The situation for optical applications is completely different. Usually, the substrates are not adapted to the fabrication technology. Instead, the size and thickness is defined by the application. This leads to a huge variety of substrates upon which lithography has to be performed, ranging from glass fibers (e.g., lithographic fabrication of gratings on the fiber core, or lens generation on the fiber end surface) on the "small" end to bulk large glass substrates for optical elements with extreme planarity requirements on the "big" end. All such substrates have to be mounted and handled in the lithography tool independent of size, thickness, and weight, with ultrahigh precision to meet the accuracy demands.

In most cases, technical solutions exist to overcome this obstacle. However, in many cases, this would require special substrate holders and/or sample stages for the lithography tool. Thus, special solutions for such substrates would become very expensive. In the case of electron-beam lithography, there are additional limitations imposed by the size of the vacuum chambers, which further add to the problem.

An alternative solution would be the use of standardized substrates also for optical applications. As a necessary precondition, this requires the early consideration of available substrate size and thickness values already in the definition phase for the function and parameters of the optical element.

Shape and Orientation of the Pattern

Most lithography tools are nowadays adapted for pattern writing in Cartesian coordinates. (There are a few exceptions where special sample stages for polar-coordinate writing [12] or special writing strategies for overcoming the Cartesian limitations [13] were developed.) The Cartesian approach is based on the typical layouts in semiconductor fabrication, where the exposure patterns are usually rectangularly shaped, with borders parallel to the coordinate axes. Only occasionally do slanted lines with 45° tilt occur in such patterns, if at all. Figure 4b shows an example.

Again, the situation is completely different for optical applications. Many applications require patterns with arbitrary tilt, such as linear gratings with varying orientation, or radial gratings for encoding wheels. Moreover, in optics, often curved shapes are employed (e.g., for waveguides, fresnel zone plates, or focusing gratings).

For such arbitrarily slanted or curved pattern emerges the question of what is the best exposure strategy. Figure 9 illustrates this problem for the example of one zone of a binary fresnel zone lens showing some different strategies for exposure with circularly or rectangularly shaped beams. One can see that the number of shots is considerably different for the cases shown, which will result in different exposure times. On the other hand, there is also a difference in

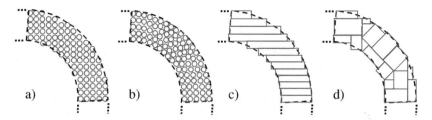

Fig. 9. Different strategies for exposure of one zone of a fresnel zone lens with circularly [(a) and (b)] or rectangularly [(c) and (d)] shaped beams.

the approximation quality of the curved border with the shots, which might influence the optical quality of the pattern.

Whether or not such an interaction between exposure strategy and optical function becomes significant depends on the optical function of the element. For example, the positioning of the individual shots on a uniform grid in Figures 9a and 9c is probably insignificant in case of a fresnel zone lens. If the same strategy is applied for a curved or slanted grating, however, such a regular shot grid may result in undesired diffraction orders, and avoiding such a regular shot grid according to Figure 9b or 9d will result in better optical quality. This shows that there is no general rule for best exposure strategy in microoptics. The effects have to be considered and discussed depending on the specific optical function of the pattern.

Relation Between Function of the Fabricated Pattern and Fabrication Costs

In many cases, an optical application or function can be realized with different elements. For example, fresnel lenses, gratings, and similar elements can be realized as binary, multilevel, or continuous profiles. The choice of the profile determines the optical efficiency of the element, but it also influences considerably the fabrication costs (which is an important issue because lithography is usually not an inexpensive technology). Thus, there is no general rule for the choice of profile. Instead, for each specific application, an agreement between customer and manufacturer is necessary to find the best solution.

Tool Performance

For the evaluation of lithography tools in the semiconductor fabrication, a number of rather standardized tests has emerged. For example, there are tests on the following:

- Placement accuracy: accuracy of placement (positioning) over the entire substrate
- Butting accuracy: local position deviation between two patterns which abut on each other

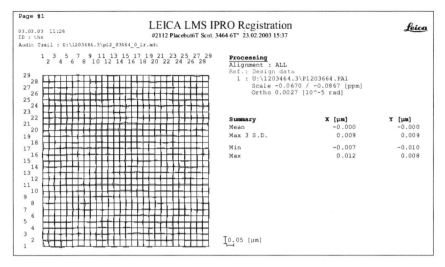

Fig. 10. Placement accuracy of Leica SB350MW e-beam writer for a 140-mm × 140-mm area; the 3σ-values are less than 10 nm.

- Linearity: realized versus nominal pattern width for a predefined range (e.g., 500 nm to 5 μm)

As an example, Figure 10 shows the results of a placement accuracy test for a Leica SB350MW shaped e-beam tool for a 140-mm × 140-mm measurement area. The maximum deviation is 12 nm and the 3σ-values are 9 nm in the x and y direction, respectively, which is a result well qualified for maskmaking for the 90-nm technology node.

For optical applications, there is the question of whether such a lithography tool characterization is sufficient. In order to answer this question, we consider the following example of two gratings fabricated by electron-beam direct writing (Fig. 11). The gratings were fabricated simultaneously with identical parameters, the only variation being the grating period, which differs by 0.99 nm. After fabrication, the optical quality was investigated by measuring the diffraction spectrum in Littrow mounting. As can be seen in Fig. 11, the optical quality of the gratings differs considerably. For a period of 342.21 nm, noticeable grating ghosts are obtained, whereas the 341.22-nm period grating shows a nearly perfect spectrum.

This effect depends on whether or not the grating period is well adapted to the lithography tool's grid of addressable positions. In the example, there is a misadaption for the grating period of 342.21 nm (left plot in Fig. 11), which results in systematic shifts (position deviations) of individual grating lines and, thus, produces the grating ghosts experimentally obtained. For the grating period of 341.22 nm, there is a good adaption to the e-beam writer's address grid and the grating ghosts are avoided.

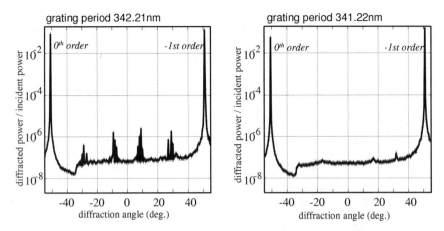

Fig. 11. Diffraction spectrum of two gratings with slightly different grating periods measured in Littrow mounting. For a period of 342.21 nm, noticeable grating ghosts are obtained, whereas the 341.22-nm period grating shows a nearly perfect spectrum.

The systematic position shifts mentioned are less than or equal to 1.25 nm, which is half of the address grid size of the lithography tool used in this example (Leica LION LV1 e-beam writer). Thus, the effect is rather small and would not become obvious in a placement accuracy test, as shown in Fig. 10. Such an effect is rather unimportant for semiconductor fabrication as long as the 3σ-values for position accuracy are in the specified range. However, for optical applications, this effect is obviously significant.

The example, therefore, shows that the requirements to the lithography tool are different between microoptics and semiconductor fabrication and that standard test procedures from the semiconductor domain to evaluate a lithography tool are not in any case sufficient for microoptics. Furthermore, it is demonstrated that the optimum performance of the fabricated optical elements requires an adaptation of the properties of the element and the lithographic technology, which can be achieved by a close partnership between the user and the supplier of the lithography tool.

4 Summary

Lithography is a powerful and widely used technology for the fabrication of optical elements. Optics benefits at this point from the continuous development and improvement of lithography tools, forced by the increasing demands of semiconductor fabrication.

This trend will certainly continue in the future, but lithography tools will be adapted more and more to the special demands of semiconductor fabrication and will partially lose its versatility. This causes challenges for optical applications, because the demands of optics on lithography are considerably different compared to semiconductor fabrication. In some cases, the resulting obstacles can be overcome by a close collaboration between supplier and user of the lithography tools and the customer (i.e., the end user of the optical elements). Thus, the future development of lithography will result in prospects as well as challenges for optics.

References

1. Brodie, I., and Muray, J.J., *The Physics of Micro/Nano–Fabrication*, Plenum Publishers, New York (1993).
2. Rai-Choudhury, P. (ed.), *Handbook of Microlithography, Micromachining, and Microfabrication*, SPIE Press, Bellingham, WA (1997).
3. Gale, M.T., Rossi, M., Pedersen, J., and Schuetz, H., Fabrication of continuous-relief micro-optical elements by direct laser writing in photoresists, Opt. Eng. 33, 3556–3566 (1994).
4. Herzig, H.P., Schilling, A., Stauffer, L., Vokinger, U., and Rossi, M., Efficient beamshaping of high-power diode lasers using micro-optics, Proc. SPIE 4437, 134–141 (2001).
5. Brünger, W., Kley, E.-B., Schnabel, B., Stolberg, I., Zierbock, M., and Plontke, R., Low energy lithography; energy control and variable energy exposure, Microelectron. Eng. 27, 135–138 (1995).
6. Yasin, S., Mumtaz, A., Hasko, D.G., Carecenac, F., and Ahmed, H., Characterisation of the ultrasonic development process in UVIII resist, Microelectron. Eng. 53, 471–474 (2000).
7. Yamazaki, K., Saifullah, M.S., Namatsu, H., and Kurihara, K., Sub-10-nm electron-beam lithography with sub-10-nm overlay accuracy, Proc. SPIE 3997, 458–466 (2000)
8. Lee, S.H. (ed.), *Diffractive and Miniaturized Optics*, SPIE Critical Reviews Vol. CR49, SPIE Press, Bellingham, WA (1993).
9. Kley, E.-B., Continuous profile writing by electron and optical lithography, Microelectron. Eng. 34, 261–298 (1997).
10. Sinzinger, S., and Jahns, J. (eds.), *Microoptics*, 2nd ed., Wiley–VCH, Weinheim (2003).
11. Borelli, N.F. *Microoptics Technology*, Marcel Dekker, New York (1999).
12. Poleshchuck, A.G., Korolkov, V.P., Cherkashin, V.V., Reichelt, S., and Burge, J.H., Polar-coordinate laser writing systems: Error analysis of fabricated DOEs, Proc. SPIE 4440, 161–172 (2001).
13. Schnabel, B. and Kley, E.-B., Evaluation and suppression of systematic errors in optical subwavelength gratings, Proc. SPIE 4231, 132–137 (2000).

Modeling of Free-Space Microoptics

Uwe D. Zeitner, Peter Schreiber, and Wolfgang Karthe

Fraunhofer-Institut, Angewandte Optik und Feinmechanik, Albert-Einstein-Str. 7, 07745 Jena, Germany
zeitner@iof.fraunhofer.de

1 Introduction

In the last decade, the use of microoptical elements such as lens arrays, single microlenses, beam splitters, and beam shapers has been constantly increased. Such microoptical elements can provide new optical functions unknown in conventional opticals. However, in most cases these elements are only part of a complex optical system containing a number of different additional optical elements. To understand the response of the whole system, the design and modeling methods for the conventional optics and the microoptics have to fit together. In this chapter, we will look at the special effects connected with the use of microoptical elements and how they can be adequately modeled. We restrict ourselves here to the consideration of free-space microoptical systems and, thus, the extensive field of integrated microoptics is beyond the scope of this chapter.

In Section 2, we summarize the consequences of miniaturization of optical elements on the overall system concept. Section 3 compares the commonly used modeling techniques, and in Section 4, some example systems are discussed.

2 Consequences of Miniaturization

A huge number of applications may potentially benefit from the use of optical techniques. However, this potential gain is often only accessible if the optics has a sufficient small size and reasonable price. Conventional optics often does not match these two important points. The scaling of component and system size down to smaller dimensions can open the way to a larger field of applications. This is a consequence of the fact that the transition from macrooptics to microoptics is not simply a scaling of size. The specific fabrication techniques developed for microoptical components offer a wide flexibility for the realization of optical functions not known from conventional optics. A popular

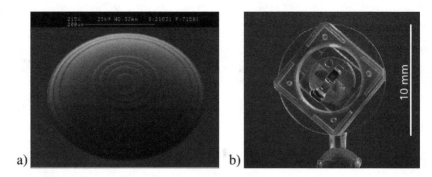

Fig. 1. (a) Hybrid microoptical element combining a refractive lens and a diffractive beam shaper. (b) Injection molded optical block of a sensor head combining nine optical elements.

example is the computer generated hologram [1,2]. Furthermore, there is also a strong impact of microoptics on overall system concepts. Proper combinations of fabrication techniques may lead to a high degree of integration. The optical tasks usually solved by different elements can be combined into one surface and the total number of elements in the system can be decreased. Figure 1a shows such an example of a diffractive beam shaper integrated into the surface of a refractive microlens which has been realized by the combination of laser lithography and reflow technology [3]. In combination with replication methods complete optical systems can be mass fabricated in a cost-effective way. As an example, Figure 1b shows an injection-molded piece combining the functions of nine optical elements.

As a result, there are many applications that only became accessible for optical solutions via microoptics. However, the scaling of size and the addition of new functionalities also affect the methods necessary for theoretical treatment in design and analysis of the systems. The reason for this mainly lies in the change of the influence of different physical effects on the optical performance during the scaling process. Although typical feature sizes are reduced, the wavelength used often remains unchanged. Especially, the strength of refraction and diffraction change in their influence on the optical function. This will be demonstrated by the simple expample of the scaling of a lens with low numerical aperture (NA). The arrangement under consideration is sketched in Figure 2a. The lens shall have a constant NA of 0.05, but its diameter and focal length are scaled down from position 1 to 3. By calculating the intensity distribution onto the optical axis along the z direction in the surrounding of the focus position, we obtain the curves shown in Figure 2b for the three different nominal focal lengths $f_1 = 1.6\,\text{mm}$, $f_2 = 0.8\,\text{mm}$, and $f_3 = 0.4\,\text{mm}$. As can be seen, the position of the highest intensity moves toward the lens as the lens size decreases. This cannot be explained if only refraction at the lens surfaces is taken into consideration. This effect is caused merely by the

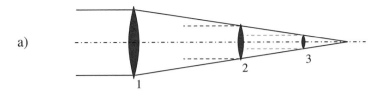

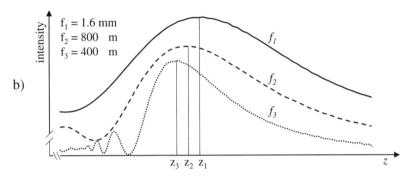

Fig. 2. (a) Arrangement for investigation of the lens-scaling effect and (b) corresponding change of focus position for a constant numerical aperture of 0.05.

increasing influence of diffraction at the lens aperture. The edge of the lens acts like the zone of a Fresnel zone plate and thus, delivers an additional optical power to the lens. This diffractive power increases as the lens diameter decreases. In most cases of practical relevance, the optical function has to be treated as a superposition of both physical effects – refraction and diffraction. The modeling techniques used for design and analysis of microoptical systems must take care of this.

3 Common Modeling Techniques

In order to get some rough ideas about the validity of the different approaches, we first have to made some systematic considerations about the crucial points in the modeling of light propagation through a microoptical system. For this purpose, we do not care about the light sources itself, but start the consideration with an electromagnetic field given at the entrance of the system. Inside of the system, we have to distinguish between the modeling of propagation through homogeneous media and the interaction of the field with elements or surfaces. The most common techniques for the modeling of these two cases are ray tracing and a wave-optics approach. However, both have different strengths and weaknesses in the two modeling regions.

3.1 Ray Tracing

The method of ray tracing is based on the solution of the eikonal equation, which can be derived directly from Maxwell's equations under the assumption of very large values of $k_0 = 2\pi/\lambda$, with λ being the wavelength [4]. As stated in Ref. [4], this approximation is justified in cases where the changes in $\mathbf{E}$ and $\mathbf{H}$ are small compared with the magnitudes of both fields over domains whose linear dimensions are of the order of λ. Consequently, the fields behave locally as a plane wave. For the modeling of optical systems, these plane waves are represented by rays whose propagation directions are given by the local directions of the pointing vector $\mathbf{S}$. In isotropic media, this direction is coincident with the direction of the k-vector of the local plane wave $\mathbf{E}_l$, given by

$$\mathbf{E}_l = \mathbf{A} \exp\{i\mathbf{kr}\}, \tag{1}$$

with $\mathbf{A}$ being the local field amplitude. For the sake of simplicity, we restrict ourselves in the following to the case of isotropic media and, thus, the consideration of the k-vector is sufficient. The light rays are, therefore, orthogonal to the geometrical wave fronts.

In the case of propagation through homogeneous media, the direction of the k-vector remains unchanged and the position of the ray after a certain propagation distance can be calculated by simple geometrical considerations. If absorption or amplification is neglected, the energy carried by the ray remains unchanged, too.

In the surrounding of inhomogeneties in optical systems, the directions of the k-vectors at the different positions on the wave front changes according to the certain form of the inhomogeneity. For the case of highest importance, which is the steplike change of the refractive index at a smooth surface separating two homogeneous media, the behavior of the k-vector can be either derived directly from the eikonal equation or by assuming local plane interfaces and using the Fresnel equations.[1] The advantage of the second approach is the possibility to include energy changes due to Fresnel losses. For the change of the ray direction, we obtain, in both cases, the laws of refraction and reflection:

$$n_i \sin \theta_i = \text{const.}, \tag{2}$$

$$\theta_1 = \pi - \theta_2, \tag{3}$$

respectively, with θ_i being the angle between the ray and the local surface normal in region i.

Many optical functions in microoptics are realized by microstructured surfaces. These have naturally strong surface profile variations and cannot be

---

[1] For a number of different smooth inhomogeneities, the behavior of the rays also follows directly from the eikonal equation. These cases will not be considered here in more detail because they are of minor significance for the understanding of the role of ray tracing for the modeling of microoptics.

considered as local plane interfaces. Furthermore, the fields interacting with such elements can also not be considered to have a plane wave front locally. Typical examples of such elements are gratings and diffractive structures. Usually, the method of ray tracing breaks down at such inhomogeneities. Of course, there are special cases in which a new direction of the k-vector after interaction with the surface can be calculated by a proper model. An example is a structure for which the local inhomogeneity can be approximated by a local periodic structure. Then, the direction of the k-vector follows from the known grating equation according to

$$n_2 \sin \theta_2 - n_1 \sin \theta_1 = \frac{m\lambda}{p} \tag{4}$$

for the one-dimensional case, with m being the diffraction order and p the local period. However, even for this case of local periodicity, the new wave front is represented by a number of plane waves or diffraction orders having different propagation directions or k-vectors. In some cases, only one direction is of interest for the further propagation through the system. Furthermore, detailed information about the distribution of energy between the diffraction orders cannot be obtained by this simple approach. In addition, for certain elements such as computer-generated holograms, a local periodicity is often not existent and, as a consequence, a change of the k-vector is hard to calculate.

Although in most cases of practical interest in conventional optics the assumption made during derivation of the eikonal equation mentioned at the beginning of this section is completely adequate in the field of microoptics, naturally the number of cases where it fails increase. In general, one can state that ray tracing is approaching difficulties in cases where strong local changes in the field amplitude and phase occur and thus, fields and surfaces can no longer be considered a local plane.

3.2 Wave-Optics Representation

The validity problems of ray tracing mainly arise from the fact that diffraction phenomena during propagation of the fields are not considered for the single rays. To include this, one must use a propagation operator for fields other than the simple geometrical consideration of ray directions. Examples are the well-known propagation operators for wave propagation in homogeneous media. The most popular ones are the angular spectrum operator and the Fresnel wave-propagation formula [5]. These operators are used to calculate the evolution of the spatial dependency of a spatial field component $U(x, y, z)$ as it propagates a certain distance Δz through the homogeneous medium. It has to be stated clearly here that these operators are not restricted to scalar considerations, but can easily used for vectorial field treatments. For most practical cases of $U(x, y, z)$, these formulas cannot be solved in a closed analytical form and numerical methods have to be applied. For this purpose, the complex field amplitude is usually discretized on sample points located ray tracing: wave-optics:

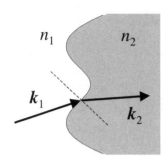
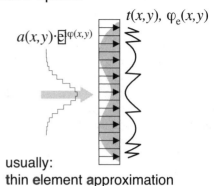

usually:
thin element approximation

Fig. 3. Different methods for representation of fields and surfaces or elements for ray-tracing and wave-optical approaches.

on a regular mesh. The resulting data volume is much larger compared with the number of parameters required for a ray-tracing approach, because one single ray at a certain position also contains information in its k-vector about the surrounding wave-front area (see Fig. 3). Often, powerful computational equipment is needed to handle this required data volume and perform the wave-optical analysis with sufficient accuracy.

One more general problem for the wave-optical approach is the treatment of the interaction between fields and inhomogeneities. There is no general method existent and a separate model for each surface type has to be used. For very few cases, a rigorous or even analytic expression exists. The best known example is the interaction of plane waves with plane surfaces expressed by Fresnels formulas. The interaction of plane waves with periodic structures can be handled, for instance, by so-called Fourier-expansion methods in a rigorous manner [6]. However, for most other surface types, certain more or less serious approximations have to be made in order to be able to handle the interaction problem. A very simple and handy method is the so-called thin-element approximation. In this approximation, the optical element under consideration is mathematically enclosed by two planes which itself are oriented perpendicular to the optical axis which is directed along the z axis. Then the optical path between the two planes is calculated along the z direction and converted into a phase function $\varphi_e(x, y)$ for each (x, y) position. Absorption is included by a proper real amplitude $t(x, y)$. As a result, one obtains the complex transmission function

$$T(x, y) = t(x, y)e^{j\varphi_e(x,y)}, \tag{5}$$

and the transmission of $U(x, y, z)$ through the element can be expressed by

$$U(x, y, z_+) = T(x, y)U(x, y, z_-), \tag{6}$$

with z_- and z_+ being the positions directly in front and directly behind the element, respectively. Equation (6) can easily be inverted, which is the big advantage of the thin-element approach, especially in design procedures. However, it becomes obvious that diffraction effects and position changes of wavefront parts during propagation inside the element are completely neglected. Thus, in most cases, the thin-element approximation represents even more of a simplification than the ray-tracing approach. In the examples section of this chapter we will see that by a proper extension of the thin-element approximation, this drawback can be overcome; however, this is usually achieved by an enormous increase of computational power.

Other modern techniques try to tackle the overall analysis of optical systems by a combination of wave-propagation methods outside inhomogeneous regions and a ray-tracing approach for the interaction at surfaces. Examples are the so-called local-plane-wave and local-plane-surface approximations [7,8]. The numerical difficulty in these methods is the proper conversion between the sampled complex amplitude and a ray-based representation of the fields without loss of accuracy and introduction of numerical artifacts. Other methods try to include the effects of surfaces and elements by tracing a bundle of test rays through inhomogeneous regions and calculate effective propagation lengths, phase aberrations, and magnification factors, which then are combined to an effective transmission operator and applied to the incident field $U(x, y, z_-)$. All of these methods have their own validity range and the user has to choose the parameters of the simulation very carefully. Often, only direct comparisons between different techniques show failures or a lack of accuracy.

3.3 Strengths of Both Methods

We briefly summarize the strengths of the two approaches discussed in the preceding subsections.

Ray-based approaches have their strengths in the following points:

- Interaction with smooth surfaces
- Relatively low number of parameters required for modeling of wave fronts, resulting in the following:
  - Simple merit function definition for imaging applications
  - Fast calculation and optimization
  - Fast tolerancing

On the other hand, the wave-optics representation of fields and components by complex amplitudes has its strength in the following cases:

- Consideration of diffraction and interference (important especially for non-imaging systems)
- Treatment of strongly modulated and arbitrary phase and amplitude profiles

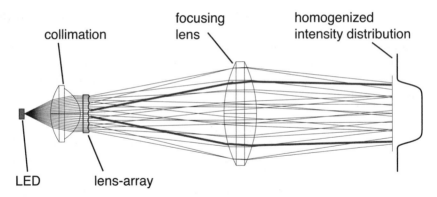

Fig. 4. Sketch of the optical arrangement for homogenization of a LED-illumination. The light going through each lens of the array is equally distributed over the whole illuminated area.

From the above discussion of the different methods, it becomes clear that for the cases of microoptics modeling and design, a clear choice between either of the approaches is far from being wise. Instead both methods are needed in parallel. The particular choice of the proper modeling technique for a certain part of a microoptical system is not only influenced by the system itself but also by the boundary conditions of the application such as required resolution, efficiency, and so forth.

4 Examples of Modeling for Microoptical Systems

The following examples are chosen to demonstrate some differences of the discussed modeling methods for microoptical systems design.

4.1 Homogenization of LED Illumination

The use of microlens arrays in optical systems for homogenous illumination of line foci or rectangular areas is a well-known technique (see, e.g., Ref. [9]). Figure 4 shows a sketch of a corresponding optical arrangement containing a light source with collimation, the lens array for homogenization, and a focusing lens performing the superposition of the light coming from different lenses in the array on the screen. The working principle is explained in detail, for example, in Ref. [10]. Here, we will focus on the modeling of such systems for use with light emitting diodes (LEDs) as the light source. In contrast to laser sources, LEDs have very different characteristics of the emitted light fields, which are the result of the completely incoherent light generation in an extended spatial region of typically some $100\,\mu m \times 100\,\mu m$ in area. The spectral bandwidth of the emitted light is usually $\Delta\lambda_{\mathrm{FWHM}} = 20$ to $100\,nm$. Such

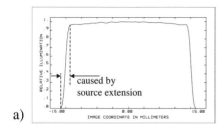
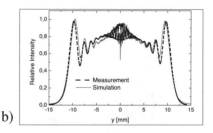

a)   b)

Fig. 5. Homogenization of an LED source with a single lens array: (a) ray-tracing analysis and (b) wave-optical analysis together with measurement results.

sources can usually be modeled well by ray-tracing methods. Here, we used a commercial ray-tracing program for estimation of overall systems parameters such as collimation lens diameter $d_c = 7.5\,\text{mm}$ and focal length $f_c = 5\,\text{mm}$, size of the lenses in the array ($d_a = 0.2\,\text{mm}$) and their focal length $f_a = 0.9\,\text{mm}$, and the focal length $f_f = 100\,\text{mm}$ of the focusing lens. The size of the LED emitter is $1.2 \times 1.2\,\text{mm}^2$. Figure 5a shows a cross section of the calculated intensity distribution on the screen. As can be seen, the ray-tracing calculation predicts a good performance of the homogenization. The slight broadening of the illuminated area is caused by the finite extension of the light source. Spherical aberrations of the array lenses result in a weak decrease of the intensity outside the center.

If we now analyze the complete optical system with a wave-optical approach and include all parameters of geometry and light source, we end up with a different result, as shown in Fig. 5b. In addition to the broadening caused by the extended source, we observe a number of other effects:

- Diffraction at the aperture of the single array lenses causing the strong low-frequency modulation of the homogenized distribution
- Diffraction at the gratinglike lens arrangement in the array causing the high-frequency modulation in the center
- Spectral bandwidth of the LED causing the finite extend of the high-frequency modulation
- Aberration of the array lenses causing the increased average intensity in the center

As can be seen by direct comparison with corresponding measurement results, these effects can be modeled in a realistic manner. However, due to the occurring diffraction effects, the overall system behavior is far away from being ideal. This can be changed by a simple extension of the optical arrangement: The single lens array is replaced by two identical lens arrays spaced at exactly their focal length f_a as depicted in Figure 6a. The result of the wave-optical analysis and the corresponding measured intensity distribution shown in Figure 6b is now much closer to the ideal homogenization. The source-extension broadening is compensated as well as the disturbing effects of diffraction at

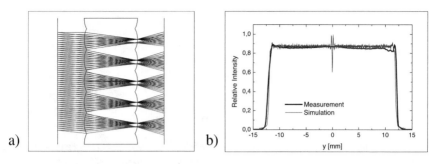

Fig. 6. (a) Tandem-arrangement of two identical arrays and (b) wave-optical analysis of the corresponding homogenization behavior together with measurement results.

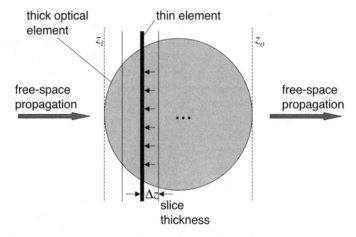

Fig. 7. Sketch of a method suitable for the modeling of wave propagation through thick elements.

the aperture of the array lenses. The influence of the lens aberrations is much weaker than for the single-lens-array system. For this arrangement also the ray-tracing calculation gives correct results.

4.2 Wave-Optical Propagation Through Thick Elements

As mentioned in Section 3.2, wave-optical modeling methods often have some difficulties in describing the propagation through elements or surfaces with a significant extension in propagation direction. In many cases, the simple thin-element approximation is not appropriate and gives incorrect results. However, this approximation can be extended to the so-called wave-propagation method (WPM) [11]. The main idea behind this method is to consider the thick optical element as a stack of different thin elements with thickness Δz. These thin slices can then be successively modeled according to Eq. (6) combined with a

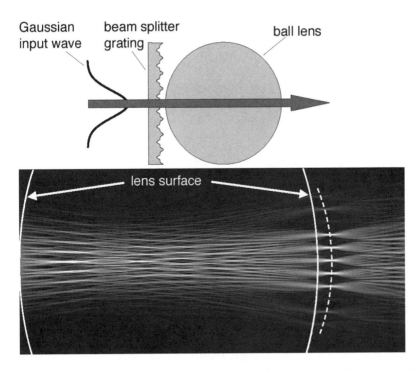

Fig. 8. Cross section of the calculated amplitude distribution for the propagation through a ball lens. The Gaussian input wave is transmitted through a 1 to 5 beam-splitter grating before entering the lens.

propagation step along a distance Δz. The method is sketched in Figure 7. An important point which has to be included in the propagation steps is the consideration of the different effective wavelengths $\lambda_{\text{eff}} = \lambda_0/n$ inside and outside the element. If the step length Δz is chosen small enough (e.g., $\Delta z < \lambda$), this can be performed by dividing the current wave front into a part completely outside the element and a part just entering the element during the step or already being inside the element. The two wave-front parts are then propagated separately along Δz with the respective λ_{eff} and combined afterward yielding the new wave front at the end of the current step. By successive application of this method, the incident wave can be propagated through the whole element. Besides the fact that a rigorous propagation operator is used and therefore the WPM is a powerful extension of purely paraxial propagation methods, there are still limitations in its validity at extreme incidence angles. This is because the boundary conditions for the vectorial field components at surfaces are usually not considered in the stepwise propagation procedure. The field components are, instead, treated as if they do not couple with each other.

As an example, Figure 8 shows the focusing of a Gaussian beam transmitted through a 1 to 5 beam splitter with a ball lens of 1 mm diameter. Depicted is the cross section of the amplitude distribution along the z-axis. The beam splitter grating is considered to be a thin element and the lens is modeled with the WPM. It can be seen that the amplitude distribution changes significantly during the propagation inside the element due to diffraction and interference effects. Also an image field curvature is observable (dashed curve) which is a non-paraxial effect. These effects could not be handled adequately with only the thin element approximation.

5 Conclusion

Due to today's fast development, microoptic technologies gain attention for more and more applications. Optical systems, in general, tend to become smaller, less expensive, and/or more efficient. Such optical systems containing elements with feature sizes in the range far below 100 μm often require different design tools or at least combinations of tools are necessary. In the present chapter, we try to give a short overview about some of the most common techniques used in microoptics design which are based on ray-tracing and wave-optics approaches. Both methods have different strengths and shortcomings. Unfortunately, sometimes it is difficult to notice a failure of a specific modeling technique. Definite, easy usable, and always applicable rules of thumb for the method of choice for a particular design problem cannot be given. Microoptic system design often requires a certain amount of experience with the optical and numerical problems which may occur. However, as we have shown by some examples, most of the difficulties of a particular modeling technique can be overcome by an appropriate extension of the method or combinations of different approaches.

Acknowledgments

The authors would like to thank their colleagues from the department Microoptics of Fraunhofer IOF and the microlithography group of E.-B. Kley from IAP, University Jena for contributions and fruitful discussions.

References

1. Kress, B. and Meyrueis, P., *Digital Diffractive Optics*, John Wiley & Sons, Chichester (2000).
2. Soifer, V., *Methods for Computer Design of Diffractive Optical Elements*, John Wiley & Sons, New York (2002).
3. Zeitner, U. and Dannberg, P., Double-sided hybrid microoptical elements combining functions of multistage optical systems, in *Lithographic and Micromachining Techniques for Optical Component Fabrication*, SPIE, Bellingham, WA (2001).

4. Born, M. and Wolf, E., *Principles of Optics*, 7th (expanded) ed., Cambridge University Press, New York (1999).
5. Goodman, J.W., *Introduction to Fourier Optics*, 2nd ed., McGraw-Hill, New York (1996).
6. Turunen, J., Diffraction theory of microrelief gratings, in *Micro-optics: Elements, Systems, and Applications*, edited by H. Herzig, Taylor & Francis, London (1997).
7. Pfeil, A. v. and Wyrowski, F., Wave-optical structure design with the local plane-interface approximation, J. Mod. Optics 47, 2335–2350 (2000).
8. Pfeil, A. v., Wyrowski, F., Drauschke, A. and Aagedal, H., Analysis of optical elements with the local plane-interface approximation, Appl. Optics 39, 3304–3313 (2000).
9. Bagdasarov, A., Size calculation of lens-array illumination systems, Sov. J. Opt. Technol. 44, 65–66 (1977).
10. Büttner, A. and Zeitner, U.D., Wave optical analysis of light-emitting diode beam shaping using microlens arrays, Opt. Eng. 41, 2393–2401 (2002).
11. Brenner, K.-H. and Singer, W., Light propagation through microlenses: a new simulation method, Appl. Optics 32, 4984–4988 (1993).

Modeling of Photonic Waveguides and Circuits

Reinhard März

Infineon Technologies AG, Corporate Research, Otto-Hahn-Ring 6, D-81739
München, Germany
Reinhard.Maerz@infineon.com

The first ideas of photonic waveguiding on the basis of total internal reflection date more than one century back. Glass fibers with losses of about 20 dB/km – the proof of feasibilty from the telecommunication point of view – was demonstrated in the early 1970s, the idea of an integrated optics (i.e., of photonic circuits on the basis of chip technology) was born at the same time [1]. The development of optical long-haul transmission lines for telecommunications and, in recent years, of optical transparent optical networks acted as the main technology driver of photonic waveguides and circuits. Driven by the ongoing explosion of the Internet, the development of optical low-cost components will be crucial for the future development of the telecommunication and data communications markets. With the increasing maturity of the photonic technologies, an increasing number of new applications such as optical frame-to-frame and board-to-board interconnections and sensors for metrology, medical and biochemistry become feasible.

The complexity of photonic circuits, which is indicated by the number of built-in devices, increased significantly during the last years. For example, reconfigurable optical ADD/DROP multiplexers (OADMs), which allow for adding and/or dropping one or several wavelength division multiplexing (WDM) channels, were integrated on a single chip. The impact of photonic integration becomes obvious if one considers that such devices replace one or more plug-in modules in a rack when configured in a conventional, hybrid way. Although these developments remain crucial for the provision of a further increasing bandwidth inside the world-wide communication networks, especially for the more cost-driven segments such as metro networks and campus backbones, the degree of integration is (and will be) small in comparison to that of mainstream microelectronic circuits.

Many of the isolated devices are, in contrast to those of microelectronics, of high complexity. For example, the phase shifter of an advanced arrayed waveguide gratings (AWGs) being able to multiplex and demultiplex 40 (and more) wavelength (WDM) channels consists of several hundred waveguides whose lengths exhibit tolerances of only a few nanometers.

Most photonic waveguide devices and the circuits built up from them operate in an essentially analog mode (i.e., the operating point varies continuously with the device parameters, which, in turn, are sensitively affected by the fabrication process).

For a long time, optical fibers as well as integrated photonic waveguides and circuits were realized using low-contrast refractive index profiles. The typical refractive index contrast at a dielectric interface is $|\triangle n| < 10^{-2}$. This layout strategy was mainly driven by the ease of fabrication (i.e., by keeping the effect of surface roughness on the insertion loss and the operation of devices small). Unfortunately, low-index-contrast waveguides require radii of curvature in the order of centimeters and turned out to be one of the greatest obstacles in order to reduce the chip sizes and thus the unit cost of integrated photonic devices and circuits. Consequently, we observe currently a clear trend to waveguides and photonic circuits based on a high refractive-index contrast $|\triangle n| = 0.01$ to 2.5. In comparison to the classical layout, the new devices and circuits exhibit a chip size reduced by orders of magnitude. Since the reduction in size is accompanied by a significant increase of birefringence, the need for advanced modeling tools that take into account the analysis of vectorial effects becomes an urgent topic.

1 Computer-Aided Engineering

To illustrate the modeling environment of photonics, we will examine the complete scenario of computer-aided engineering (CAE) of a chip technology. The essential tasks of computer-aided engineering (see Fig. 1) are currently modeling and preparing layouts for the fabrication of components.

- *Process Modeling*: covering both the numerical analysis of the fabrication processes and equipment used for the fabrication
- *Device Modeling*: the low-level analysis of the optical field inside an optical component as well as the analysis of the interaction with the electrical field in optoelectronic components
- *Circuit Modeling*: the higher-level analysis of the operation of a set of devices connected within a photonic circuit

The modeling path starts from the physical layout of a component (i.e., from a model of the device defining both geometry and materials). In a first step, the process modeling will yield the device description (e.g., a directional coupler), providing its geometry and materials (e.g., the refractive index profile). The subsequent device modeling (e.g., the beam-propagation method) analyzes the functional behavior of the device starting from this description. It forecasts a set of device parameters (for microelectronics: SPICE parameters) describing the operation of the device within a circuit or any other composite structure. The layout path (CAD), in contrast, starts from the system design. The physical description (i.e., the true mask data of the chip) is worked out

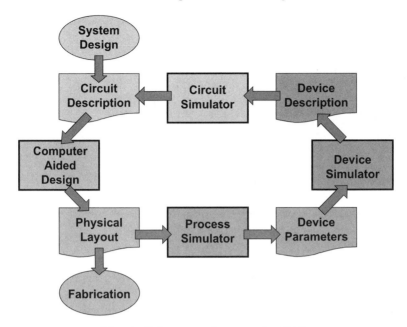

Fig. 1. Software environment for CAE.

from the system design by using one – for microelectronics more – intermediate level of circuit descriptions. It should be noted that these two paths of CAE are to a large extent independent of each other.

The acceptance of design and modeling as an essential part of the photonical development process is no longer at issue. Especially, the product development of photonic devices and circuits is increasingly accompanied by an extensive modeling of their properties in order to reduce the number of expensive and time-consuming development cycles. In parallel, the design engineers are faced with the following, increasingly challenging modeling tasks:

- *Visualization*
  The visualization of the optical field inside an integrated optical component irrespective of the accuracy of the underlying results was the first goal of integrated optical design modeling, especially of the beam propagation method developed in the early 1980s. The visualization of vectorial fields and true phase fronts is a task that is still ongoing.
- *Numerical Experimentation*
  Device modeling, in particular the analysis of propagation in either time or frequency domain, can be regarded as a computational experiment where each run corresponds to one experiment. Although scanning near-field microscopy allows one to trace the optical field inside a device, representing a quantum leap in optical imaging, it is still much easier to analyze the

effect of specific stimuli on the optical field inside a hybrid device or an integrated optical chip numerically.

- *Backtracking*
  In order to debug devices and circuits, there is an increasing interest to trace back the operation of a fabricated device in order to assess the influence of manufacturing deviations on the operation by simulating the truly manufactured devices with all of their deviations ("technology shortcut"). Simulators used for backtracking must be able to resolve local deviations, which are typically one or more orders of magnitude smaller than the device under investigation. Examples of such deviations are local defects, displacements, and inhomogeneities. In contrast to the more qualitative modeling tasks formulated above, backtracking relies on modeling tools which allow one to investigate multiscale structures with high, at least relative, accuracy.

- *Tolerancing*
  The reduction of fabrication tolerances is one of the primary tasks within the design of analog devices and circuits. The requirements for tolerancing include those of backtracking. In addition, it becomes crucial to offer fast solvers – tolerancing is usually based on a large number of computations – which allow for the interfacing of multipurpose statistical tolerancing tools already in use in industrial application. Tolerancing is not limited to the low-level device modeling but also applied to find the tolerances of the parameters determining the operation of composite photonic devices and circuits.

- *Parameter Forecasting*
  Parameter forecasting represents the most challenging discipline of modeling. Driven by the economic impact, companies all over the world aim at a first-pass design of their products. A successful first-pass design of an analog device relies on solvers exhibiting a high absolute accuracy which can be – at least a posteriori – controlled. The availability of highly accurate input parameters for the device modeling (i.e., a sufficiently detailed knowledge of the fabrication processes) usually form the crucial obstacle preventing a successful implementation of a first-pass design.

2 Device Modeling

Device modeling covers a set of tools and methods which help to analyze the behavior of active and passive devices on lowest level (i.e., by computing the optical field inside the device). It includes the eigenmode analysis delivering input parameters for the subsequent circuit analysis and/or the layout of optical systems for fiber-to-chip coupling as well as propagation-oriented methods used to study the behavior of a device by computational experiments (e.g., by the excitation with specific stimuli).

The increasing use of device modeling in an industrial environment still generates a growing demand for solvers which are able to analyze the finally fabricated waveguide structures with a sufficiently high accuracy. In particular, this includes the following:

- The computation of modal gain and loss for inhomogeneously pumped lasers and amplifiers
- The analysis of leaky waveguide structures such as curved waveguides or waveguides separated by a buffer from a high-refractive-index substrate
- The derivation of design rules for the placement of metallic structures (e.g., heaters)
- The more accurate computation of polarization-dependent losses and propagation constants for both fibers and integrated optical devices
- The analysis of parasitic effects such as stress and heat flow on waveguides
- The computation of eigenmodes of full 3-D resonators especially for active components (e.g., vertical cavity surface emitting lasers)

Device modeling is based on a variety of mathematical techniques, including finite differences, finite elements, mode matching, method of lines, plane-wave techniques, and several techniques using equivalent sources [2–6]. The availability of feasible, more accurate descriptions of the chromatic dispersion, temperature dependence, and gain/loss for the material systems [7–11] turns also out to be crucial for the success of device modeling.

2.1 Eigenvalues of Simple and Composite Waveguide Structures

The eigenmode analysis of waveguides and waveguiding structures is the generic task of a component designer in order to get information about the discrete spectrum of eigenvalues of the waveguide structure. For a long time, the interest of designers was concentrated on the fundamental mode and/or a few excited modes. The eigenmode analysis was usually restricted to the case of power conservation; the influence of gain and loss was incorporated via confinement factors (i.e., on the level of a first-order Rayleigh–Schrödinger perturbation theory). In contrast to microstrip lines [12], optical waveguides are open (i.e., not enclosed by a metal boundary).

For visualization reasons, the numerical calculation of eigenvalues by using equations for the transverse field components is superior, since the optical waves propagating through typical waveguide structures are "nearly" transverse; that is, the transverse field components mainly determines the shape of the optical field. In addition, the calculation of eigenvalues relies typically on the magnetic field $\mathbf{H}_t$ since at optical frequencies the transverse magnetic field is continuous in the transverse plane.

The canonical eigenvalue problem delivering the relative permittivity ϵ_m and the amplitude of the vectorial field or one of its components $\mathbf{f}^{(m)}(\mathbf{r}_t)$ is given by

$$\mathcal{H}\mathbf{f}^{(m)}(\mathbf{r}_t) = \epsilon_m \mathbf{f}^{(m)}(\mathbf{r}_t), \tag{1}$$

where $\mathcal{H}$ stands for the Helmholtz operator describing the waveguide structure. The following list shows the most popular formulations of the eigenvalue problem [13–17] for different types of waveguides in comparsion.

- *Vectorial Formulation*
  The vectorial formulations of the eigenvalue problem in terms of the two transverse field components $\mathbf{h}_t^{(m)}(\mathbf{r}_t)$ and $\mathbf{e_t}^{(m)}(\mathbf{r}_t)$

$$\mathcal{H}_{\mathbf{H}}^{(t)} = \frac{1}{k_0^2}[\triangle_t + k_0^2\epsilon(\mathbf{r_t}) + (\nabla_t \ln \epsilon(\mathbf{r_t})) \times \nabla_t \times], \tag{2}$$

$$\mathcal{H}_{\mathbf{E}}^{(t)} = \frac{1}{k_0^2}[\triangle_t + k_0^2\epsilon(\mathbf{r_t}) + \nabla_t(\nabla_t \ln \epsilon(\mathbf{r_t})) \cdot] \tag{3}$$

  represent the most popular rigorous formulations of the eigenvalue problem since they always deliver the dominant field component, avoid spurious solutions, and offer computational efficiency by relying on the coupling of only two field components. Equation (2) is usually prefered since the transverse magnetic field components are continuous in the transverse plane. Nevertheless, for some numerical algorithms such as finite-element solvers based on edge elements, it is superior to solve the full 3-D vectorial Helmholtz equation and remove the spurious solutions numerically. The competing E_z/H_z formulation is only rarely applied today since it delivers the (nondominating) longitudinal field components.
- *Semivectorial Formulation*
  The semivectorial formulation of the eigenvalue problem represents an approximation for waveguide structures based on a slablike structure delivering only the dominating field component. Equation (4) is used to compute the transverse electric (TE) modes (also HE modes) whose electric field runs parallel to the layers; Equation (5) describes the transverse magnetic (TM) modes (also EH modes) whose magnetic field runs parallel to the layered waveguide structure.

$$\mathcal{H}_{\mathbf{TE}}^{(SV)} = \frac{1}{k_0^2}\left[\frac{\partial^2}{\partial x^2} + \frac{\partial^2}{\partial y^2} + k_0^2\epsilon(\mathbf{r_t}) - \frac{\partial \ln \epsilon(\mathbf{r_t})}{\partial y}\frac{\partial}{\partial y}\right], \tag{4}$$

$$\mathcal{H}_{\mathbf{TM}}^{(SV)} = \frac{1}{k_0^2}\left[\frac{\partial^2}{\partial x^2} + \frac{\partial^2}{\partial y^2} + k_0^2\epsilon(\mathbf{r_t}) - \frac{\partial \ln \epsilon(\mathbf{r_t})}{\partial x}\frac{\partial}{\partial x}\right] \tag{5}$$

  Both equations are formulated in terms of the magnetic field components $h_x^{(m)}(\mathbf{r}_t)$ and $h_y^{(m)}(\mathbf{r}_t)$, which are both continuous over the transverse plane. The inaccuracy of the computed eigenvalues and eigenmodes imported by this approximation is unknown up to now.
- *Scalar Formulation*
  The scalar formulation of the eigenvalue problem, that is, the use of the scalar Helmholtz equation

$$\mathcal{H}_S = \frac{1}{k_0^2} \left[\Delta_t + k_0^2 \epsilon(\mathbf{r}_t) \right], \tag{6}$$

fully neglects the effect of polarization. In consequence, the scalar formulation aims at the analysis of any vector component $\phi^{(m)}(\mathbf{r}_t)$ of a weakly guiding waveguide structure. Based on the computed results, the error caused by this approximation can be estimated a posteriori by regarding its results as a reference solution for a subsequent Rayleigh–Schrödinger perturbation expansion in terms of the vectorial character.

- *TE/TM Formulation for Slab Waveguides*
  Equations (7) and (8)

$$\mathcal{H}_{\mathbf{TE}} = \frac{1}{k_0^2} \left[\frac{d^2}{dx^2} + k_0^2 \epsilon(x) \right], \tag{7}$$

$$\mathcal{H}_{\mathbf{TM}} = \frac{1}{k_0^2} \left[\frac{d^2}{dx^2} + k_0^2 \epsilon(x) + \frac{d \ln \epsilon(\mathbf{x})}{dx} \frac{d}{dx} \right] \tag{8}$$

correspond to the rigorous TE and TM formulations of the eigenvalue problem for an arbitrarily layered waveguide in terms of the transverse field components $e_y^{(m)}(x)$ and $h_y^{(m)}(x)$, respectively.

Some photonic eigenvalue problems

$$\mathcal{O}\mathbf{F}^{(m)}(\mathbf{r}_t) = \left(k_0^{(m)} \right)^2 \mathbf{F}^{(m)}(\mathbf{r}_t) \tag{9}$$

are inverse; that is, the solution of the eigenvalue problem delivers the frequencies or the vacuum wavevectors $k_0^{(m)}$ for a given effective permittivity (or wavevector). The computation of the band structure of photonic crystals [18–20]

$$\mathcal{O}_H = (\nabla + i\mathbf{q}) \times \left[\frac{1}{\epsilon(\mathbf{r})} (\nabla + i\mathbf{q}) \times \right] \tag{10}$$

represents a prominent example of that kind. For lossless structures, the inverse formulation of the eigenvalue problem represents a minor issue, since the inversion of the waveguide dispersion $k_0^{(m)}(\mathbf{q})$ at a given propagation direction $(\mathbf{q}/|\mathbf{q}|)$ does not require a significant computational effort. For the complex case, however, the inverse formulation leads to physically meaningless complex eigenfrequencies which have to be converted into real values by introducing appropriate complex wavevectors for each band of the band structure. Obviously, this procedure results in an additional numerical effort.

The control over the accuracy of the eigenmodes and the associated effective permitivities is gaining importance for the application of the numerical tools in an industrial environment, especially in critical parameter regions leading to nearly degenerate eigenmodes, eigenmodes close to cutoff, and singularities in the computed fields. To roughly assess the accuracy of the eigenvalues, it is an established strategy for algorithms based on sampling such

64   Reinhard März

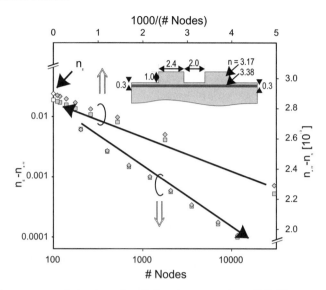

Fig. 2. Convergence behavior of a finite-element solver [21] illustrated by using the fundamental (diamonds) and first excited (boxes) eigenmode of a strip-loaded, weakly coupled directional coupler.

as finite differences or on an expansion into local or global functions such as finite-element methods or plane-wave expansions to plot the approximated eigenvalue or a derived expression over the inverse number of unknowns (i.e., over the inverse number of sampling points or functions used to approximate the eigenvalue). For the vectorial Helmholtz equation (2), Figure 2 (top/right axes) shows for illustration the convergence behavior of a finite-element solver based on linear nodal elements [21]. The approximations of the fundamental mode and the first excited mode of a strip-loaded directional coupler and the eigenvalue extrapolated from the last two computations were used to compute the extrapolated effective refractive indices n_E. The parameters of the directional coupler are shown in Figure 2. A detailed assessment of the relative numerical accuracy is based on the logarithmic deviation of the approximations from the exact eigenvalue. It requires the knowlegde of the exact eigenvalue or, as a work-around, a self-consistently computed eigenvalue of extremely high accuracy. Figure 2 (bottom/left axes) shows a log/log plot of the converging eigenvalues, where the extrapolated value n_E is used as a reference for the corresponding eigenvalue. If the reference value is insufficiently accurate in comparison to its approximations under test, a make-believe end of convergence indicated by a flattening of the convergence curve will occur. A feasible assessment of the convergence of the computed field values at the sampling points required for the layout of advanced optical systems for fiber-to-chip coupling is not yet available.

2.2 Time-Domain Propagation

Time-domain propagation methods analyze the response of a device to a spatio-temporal stimulus such as a pulse or a sequence of pulses. They directly implement the concept of computational experiments. Similar to signal processing, the time-domain propagation methods allow one to study the full spectral response of a device within a single step by examining the response of a device to an short pulse (impulse response). Obviously, the spectral resolution and the number time steps are interdependent by a Fourier transform. The analysis of highly resonant features and other phenomena which rely on eigenmodes with a small group velocity therefore requires a large number of sampling points in the time domain. The material dispersion must be taken into account via inverse convolution.

The computation of time-domain propagation is usually restricted to a certain computational window. The realization of appropriate boundary conditions (e.g., perfectly matched layers (PMLs) [22]) are crucial for the suppression of artificial reflections at the boundary. The computed results are analyzed by a numerical detector which integrates the desired quantity – usually the optical power, the Poynting vector, or a field component – over a number of sampling points.

Currently, time-domain propagation is usually implemented by using the finite-difference time-domain (FDTD) algorithm [23, 24]. The equations

$$E_z(\mathbf{r}_t, t + \Delta t) = E_z(\mathbf{r}_t, t) + \tfrac{\Delta t}{\epsilon(\mathbf{r}_t)} \left[\tfrac{\partial H_y(\mathbf{r}_t, t + \Delta t/2)}{\partial x} - \tfrac{\partial H_x(\mathbf{r}_t, t + \Delta t/2)}{\partial y} \right],$$

$$H_x(\mathbf{r}_t, t + 3\Delta t/2) = H_x(\mathbf{r}_t, t + \Delta t/2) - \Delta t \tfrac{\partial E_z(\mathbf{r}_t, t + \Delta t)}{\partial y},$$

$$H_y(\mathbf{r}_t, t + 3\Delta t/2) = H_y(\mathbf{r}_t, t + \Delta t/2) + \Delta t \tfrac{\partial E_z(\mathbf{r}_t, t + \Delta t)}{\partial x} \tag{11}$$

for the propagation of transverse electric (TE) modes and

$$H_z(\mathbf{r}_t, t + \Delta t) = H_z(\mathbf{r}_t, t) - \Delta t \left[\tfrac{\partial E_y(\mathbf{r}_t, t + \Delta t/2}{\partial x} - \tfrac{\partial E_x(\mathbf{r}_t, t + \Delta t/2}{\partial y} \right],$$

$$E_x(\mathbf{r}_t, t + 3\Delta t/2) = E_x(\mathbf{r}_t, t + \Delta t/2) + \tfrac{\Delta t}{\epsilon(\mathbf{r}_t)} \tfrac{\partial H_z(\mathbf{r}_t, t + \Delta t)}{\partial y},$$

$$E_y(\mathbf{r}_t, t + 3\Delta t/2) = E_y(\mathbf{r}_t, t + \Delta t/2) - \tfrac{\Delta t}{\epsilon(\mathbf{r}_t)} \tfrac{\partial H_z(\mathbf{r}_t, t + \Delta t)}{\partial x} \tag{12}$$

for the propagation of transverse magnetic (TM) modes show an implementation of a finite-difference scheme for the analysis of signals propagating through a general 2-D medium. Target structures for such simulations are 2-D photonic crystals where the transverse field components run parallel to holes or pillars forming the periodic structure[1] and the mapping of photonic

---

[1] Many publications on quasi-2-D photonic crystals borrow their definitions of TE and TM modes from the underlying slab waveguide; that is, the definitions of TE and TM modes are exchanged in comparison to the strict electrodynamic definitions used in here.

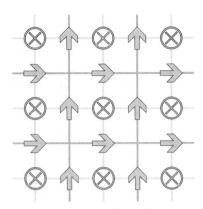

Fig. 3. Local distribution of sampling points of transverse (crosses) and the two non transverse (arrows pointing right and up) for the finite-difference time-domain method ("Yee's cell").

circuits onto an approximate 2-D equivalent in order to save computational effort.

Yee's cell (see Fig. 3) sketches the local distribution of sampling points for the field components of a full FDTD scheme, replacing all spatio-temporal differential operators by central differences. Obviously, the transverse field component (E_z for the TE case and H_z for the TM case) and other field components (H_x and H_y for the TE case and E_x and E_y for the TM case) form three different grids which are shifted relative to each other. Equations (11) and (12) show that the sampling points in the time domain for the transverse and the other field components are also shifted with respect to each other. The computational error of the FDTD is determined by the choice of appropriate spatio-temporal steps. It can be shown that the use of "magic" time steps $\Delta t = \Delta x/c = \Delta y/c$ minimizes the computational errors by annihilating the leading-order deviations.

As a practical consequence, a high spatial resolution required by multiscale geometries exhibiting tiny details forces the use of small time steps and results in a dramatic increase of the computational effort. From a physical point of view, this interdependence is a consequence of the finiteness of the velocity of light which, in turn, makes it necessary to adapt an electromagnetic field running through a finely structured medium more frequently.

2.3 Frequency-Domain Propagation

Frequency-domain propagation methods aim at the analysis of the time-harmonic behavior (i.e., of the quasi-steady state which is obtained when all transient phenomena fade away). Based on the time-domain propagation methods described above, the analysis of this state requires one to run computational experiments running over a long time. By multiplying the computed results of a simulation with the phase factor $\exp(-i\omega t)$, it is possible to visualize the true time-harmonic propagation (at the frequency under investigation).

The beam-propagation method (BPM), the most popular frequency-domain propagation scheme, starts from the (time-harmonic) Helmholtz equation

$$\left[\frac{\partial^2}{\partial z^2} + k_0^2 \mathcal{H}\right] \mathbf{F}(\mathbf{r}_t, z) = \mathbf{0} \tag{13}$$

of a z-invariant waveguide structure, where $\mathcal{H}$ describes the Helmholtz operator at any level of approximation and $\mathbf{F}(\mathbf{r}_t, z)$ is the associated vectorial field or a single field component. Within this basic scheme, it is obviously impossible to describe plane waves running perpendicular to the propagation direction. By switching to the corresponding forward Helmholtz equation, the optical field is divided into forward and backward running waves and the simulation is restricted to the analysis of the waves running into the half-space $z > 0$. By subtracting a reference plane wave running parallel to the z axis through a medium with a (proper chosen) mean refractive index $\bar{n}$, it is possible to formulate a first-order differential equation

$$\frac{\partial \mathbf{f}(\mathbf{r}_t, z)}{\partial z} = ik_0(\sqrt{\mathcal{H}} - \bar{n})\mathbf{f}(\mathbf{r}_t, z), \tag{14}$$

which governs the make-believe evolution of a make-believe amplitude in the half-space $z > 0$. The formulation of the frequency-domain propagation in terms of a slowly varying amplitude often allows for propagation steps of several wavelengths inside the material. The forward Helmholtz equation represents a linear, first-order partial differential equation whose formal solution is given by

$$\mathbf{f}(\mathbf{r}_t, z + \Delta z) = \mathcal{U}_{\text{BPM}}\mathbf{f}(\mathbf{r}_t, z), \tag{15}$$

with the BPM propagator

$$\mathcal{U}_{\text{BPM}} = \exp(ik_0(\sqrt{\mathcal{H}} - \bar{n})\Delta z). \tag{16}$$

The implementation of the corresponding computer codes takes the stimulus, propagates it over a single step through a z-invariant medium, regards the result as a new stimulus, propagates it over the next step through a modified z-invariant medium, and so on. As a consequence, the beam-propagation method describes the propagation of the optical field in a stratified medium whose boundaries run either parallel or perpendicular to the z axes. The algorithms used for beam propagation scheme are classified by the treatment of its constituent parts:

- *Helmholtz Operator*
  Driven by the restrictions of the computer infrastructure, most of the commercial BPM solvers currently use the scalar Helmholtz operator, usually in two dimensions. Vectorial solvers are still rarely applied.

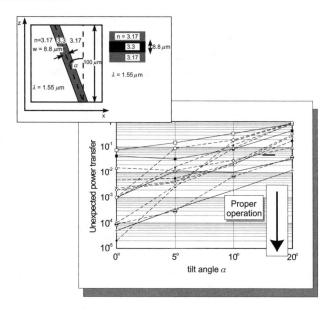

Fig. 4. "Tilted-waveguide" benchmark example showing the angular behavior of BPM algorithms (for details, see Refs. [25] and [26]).

- *BPM Propagator*

  The BPM propagator (16) can only be calculated analytically if the underlying Helmholtz operator is diagonal; that is, if the field amplitude $\mathbf{f}(\mathbf{r}_t, z)$ is expanded into the eigenmodes of the waveguide structure. Algorithms based on the eigenmode expansion require an eigenmode analysis for every new cross section and an expansion of the stimulus into the eigenmodes. They have been realized and turned out to be well suited for the behavioral modeling of devices operating on the basis a few guided modes. The numerical cost for the eigenmode expansion increases rapidly with the number of modes, the appropriate modeling of radiation is delicate.

  All other algorithms tackle the BPM propagator (16) by a selection of terms in its series expansion. The paraxial approximation, which is equivalent to the computation of the Fresnel integral, replaces the BPM propagator by the two leading terms of its Taylor expansion. It is essentially a small-angle approximation. The decreasing stability of the classical algorithms which are based on this approximation became obvious by examining the propagation of guided modes of a multimode waveguide in the framework of a benchmark example. Figure 4 indicates that the trusted range of angles for the high-order guided modes under investigation is below 10°. The results of the benchmark test indicates that the numerical errors of the radiation modes are even higher.

  The development of wide-angle approximations [27] (i.e., the development of more sophisticated expansions of the BPM propagator), was stimulated by this benchmark test. It led to algorithms which allow for the propagation of the fundamental mode in waveguides which are tilted up to angles of 45° with respect to the z axis.

- *Dielectric Interface*
  Bidirectional BPM algorithms allow for a limited treatment of reflection by computing the reflection at the dielectric interfaces at the interface between different cross sections. The reflections are handled on the level of local Fresnel reflections or by satisfying the dielectric boundary conditions in the framework of a fully vectorial beam propagation.

More sophisticated approaches of frequency-domain propagation rely also on the separation of propagation directions. However, they divide the optical field into an incoming field entering the computational window and a scattered field leaving the computational window (forever) and thus avoid defining a main propagation direction (z axis). The basic idea of such algorithms runs as follows: For lossless media, the asymptotic optical field is governed by Sommerfeld's radiation condition (precisely speaking, the Silver–Müller condition) [28]. By using this result, it is possible to divide the optical field at infinity into two parts: the scattered field leaving the "world" forever and an incoming field entering the world for the first time. By assuming homogeneity outside the computational window used for the true simulation, it is obviously possible to separate both contributions also at the boundary of the computational domain.

Obviously, the incoming field at the boundary can be used as a stimulus for simulations. The implementation of any numerical algorithm strictly relies on an effective dereflection of the boundary of the computational window. From the user's point of view, the advanced frequency-domain approaches add the following options to the classical BPM approach:

- Omnidirectional scattering created by arbitrarily shaped inhomogeneities
- Use of omnidirectional stimuli entering the computational window from all directions

Advanced refinements algorithms [29] of that type can also do without the restriction to a homogeneous background; that is, it allows for inhomogeneous structures outside the computational window. Practical examples are dielectric interfaces running through the computational window, waveguides entering the computational window, large structures such as mirrors, or a periodic background formed by a Bragg grating or a photonic crystal (s. Fig. 5).

It should be noted that the substantial increase of flexibility of the omnidirectional frequency-domain propagation methods is achieved at the expense of a significantly higher computational effort, since the algorithms rely on the computation of the rapidly varying optical field and not a slowly varying quasiamplitude as the BPM algorithms.

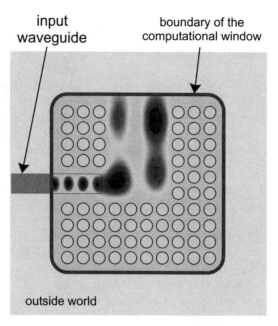

Fig. 5. Frequency-domain propagation of an optical field through a photonic crystal (courtesy of F. Schmidt, Konrad-Zuse Institut für Informationstechnik, Berlin).

3 Circuit Modeling

The device modeling presented above represents the appropriate way to analyze the operation of isolated components and small circuits by computing the optical field. The analysis of systems of larger circuits [15, 30–33] require simplifications to reduce the computational effort in space and time. However, as in microelectronics, the need to tackle the operation of increasingly complex circuits by replacing their constituent parts by a few characterizing numbers forms the primary reason to formulate the operation of circuits in simpler terms. As the classical SPICE simulation often applied in electronics, the circuits analysis in photonics relies on components which are connected by optical waveguides or fibers whose guided modes transport optical signals from device to device. Although it is not required, the interconnecting waveguides support usually only one guided mode.

The operation of optical waveguides is frequently formulated in terms of transfer matrices, a photonic equivalent to the wave-amplitude transmission matrices used for the analysis of microwave circuits [34], since this approach allows one to transport the optical field from the input side to the output side of a waveguide device by a simple matrix multiplication and thus to stack devices or partial circuits by simply stacking their constituent matrices. It should be noted that this approach can lead to numerical instabilities in the case of tunneling, in particular when the evanescent field drops dramatically inside a device.

The transfer matrix $\mathcal{U}$ describes the transfer of set of forward $(+)$ and backward $(-)$ traveling waves through a circuitry of optical devices [35]. The input and output vectors represent the amplitudes corresponding to the eigenmodes of the (usually single mode) waveguides forming the input and output ports of the circuitry. The vector of the transferred modes is given by

$$
\begin{pmatrix} \mathbf{\Phi}_{\mathrm{RIGHT}}^{(+)} \\ \mathbf{\Phi}_{\mathrm{RIGHT}}^{(-)} \end{pmatrix} = \left(\begin{array}{c|c} \mathcal{U}_{++} & \mathcal{U}_{+-} \\ \hline \mathcal{U}_{-+} & \mathcal{U}_{--} \end{array} \right) \begin{pmatrix} \mathbf{\Phi}_{\mathrm{LEFT}}^{(+)} \\ \mathbf{\Phi}_{\mathrm{LEFT}}^{(-)} \end{pmatrix}.
\tag{17}
$$

Within Equation (17), $\mathcal{U}_{++}$ describes the interaction between the forward traveling waves $(+)$ and $\mathcal{U}_{--}$ describes that between the backward traveling waves $(-)$. $\mathcal{U}_{-+}$ stands for the interaction between the two incident waves and $\mathcal{U}_{+-}$ for that between the two outgoing waves.

The optical power $P = (\mathbf{\Phi}^*, \mathcal{M}\mathbf{\Phi})$ – precisely speaking, the square of the electric or magnetic field – is conserved, if the diagonal metric tensor $\mathcal{M}$ is conserved by a transfer matrix $\mathcal{U}$, whose positive values apply to the forward traveling modes and the negative values to the backward traveling modes. The transfer matrices $\mathcal{U}$ satisfying the conservation law are designated as $U(M, M)$ of (M, M)-unitary matrices.

Usually, the transfer matrices are used to analyze the response of a circuitry to stimuli on its left-hand side $(\mathbf{\Phi}_{\mathrm{RIGHT}}^{(-)} = 0)$. Under this condition, Eq. (17) can be used to derive the generalized laws of reflection

$$
\mathbf{\Phi}_{\mathrm{LEFT}}^{(-)} = -\mathcal{U}_{--}^{-1} \mathcal{U}_{-+} \, \mathbf{\Phi}_{\mathrm{LEFT}}^{(+)},
\tag{18}
$$

which describes the backward traveling modes in terms of the incident modes. Based on that result, it is easy to derive the generalized law of transmission

$$
\mathbf{\Phi}_{\mathrm{RIGHT}}^{(+)} = \left(\mathcal{U}_{++} - \mathcal{U}_{+-} \mathcal{U}_{--}^{-1} \mathcal{U}_{-+} \right) \mathbf{\Phi}_{\mathrm{LEFT}}^{(+)},
\tag{19}
$$

which describes the forward traveling field in terms of the incident field.

For purely codirectional devices, the transfer matrix has a box-diagonal form; the propagation problem can then be reduced to one for the forward traveling modes with a transfer matrix $\mathcal{U}_{++}$. The metric tensor of the reduced problem is given by the unity matrix. From that, it can be further concluded that the propagator $\mathcal{U}_{++}$ for a lossless circuitry is a unitary operator.

For the most elementary case of two interacting modes, the transfer matrices usually have the form

$$
\mathcal{U}\pm = \begin{pmatrix} A^{(=)} & A^{(\times)} \\ \mp A^{(\times)^*} & A^{(=)^*} \end{pmatrix},
\tag{20}
$$

where the upper sign applies to the codirectional case and the lower sign to the contradirectional one. Regardless of the power conservation, each type

72   Reinhard März

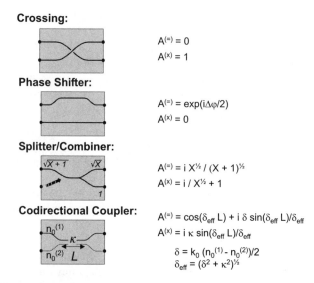

Fig. 6. Transfer matrices for elementary codirectional devices.

of transfer matrix ($\mathcal{U}_+$ and $\mathcal{U}_-$) forms a group; that is, the product of two transfer matrices of this type results in a transfer matrix of the same type.

If the optical power is conserved [i.e., if the transfer matrices belong to the groups $U(2)$ (codirectional case) or $U(1,1)$ (contradirectional case)], the transfer matrices can be made unimodular ($\det U_\pm = 1$) by an appropriate phase factor. The transfer matrix of an N-period stack is then

$$\mathcal{U}_\pm^N = U_{N-1}\left(\left|\Re\left(A^{(=)}\right)\right|\right)\mathcal{U}_\pm - U_{N-2}\left(\left|\Re\left(A^{(=)}\right)\right|\right)\mathcal{E}. \tag{21}$$

$U_N(x)$ designates the Chebyshev polynominals of the second kind. Figures 6 and 7 show some of the most elementary codirectional and contradirectional devices, respectively. Transfer matrices for more sophisticated devices are found in Refs. [32] and [36].

Emerging fields such as photonic crystals were up to now resistent to simple circuit descriptions. The description of devices affected by resonances by a few (localized) Wannier functions [37] offers a promising way to describe the operation of such devices close to resonances by a simple, tractable theory.

4 Concluding Remarks

We have taken a short tour through the world of modeling for photonic waveguides and circuits. The chapter was focused on the mainstream of modeling for passive photonic circuits (i.e., computation of eigenvalues of straight waveguides' time- and frequency-domain propagation and transfer matrix analysis

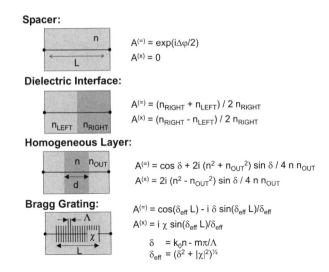

Fig. 7. Transfer matrices for elementary contradirectional devices.

of photonic circuits). Many interesting topics such as optoelectronic modeling, modeling of multimode structures, thermo-optical modeling, modeling of losses by curved waveguides and surface imperfections, and so forth were not treated here.

Photonic modeling is increasingly driven by industrial needs and, as a consequence, face new and interesting challenges such as backtracking and tolerancing which, in turn, results in a demand on controllable accuracy. The advent of high-index-contrast waveguides and circuits makes it necessary to design an increasing number of devices on the basis of a vectorial analysis.

References

1. Tamir, T. (ed.), *Guided-Wave Optoelectronics*, 3rd ed., Springer-Verlag, Berlin (1988).
2. Yamashita, E. (ed.), *Analysis Methods for Electromagnetic Wave Problems*, Artech House, Boston (1990).
3. R. Pregla, Chapter 5, this book
4. Schmidt, F., An adaptive approach to the numerical solution of Fresnel's wave equation, IEEE J. Technol. 11, 1425–1434 (1993).
5. Reed, M., Sewell, P., Benson, T.M., and Kendall, P.C., An efficient propagation algorithm for 3D optical waveguides, IEE Proc., 145, 53–58 (1998).
6. Pregla, R., von Reden, W., Hoekstra, H.J.W.M., and Baghdasaryan, H.V., Beam propagation methods, in *Photonic Devices*, Springer-Verlag, Berlin (1999).
7. Adachi, S., *Physical Properties of III-V Semiconductor Compounds*, Wiley, New York (1992).

8. Jenkins, D.W., Optical constants of $Al_xGa_{1-x}As$, J. of Appl. Phys. 68, 1848–1853 (1990).
9. Palik, E.D. (ed.), *Handbook of Optical Constants of Solids*, Academic Press, London (1985).
10. Ghosh, G. (ed.), *Handbook of Thermo-Optical Coefficients of Optical Materials With Applications*, Academic Press, London (1998).
11. Piprek, J., *Semiconductor Optoelectronic Devices*, Academic Press, London (2003).
12. Collin, R.E., *Field Theory of Guided Waves*, 2nd ed., IEEE Press, New York (1986).
13. Snyder, A.W. and Love, J.D., *Optical Waveguide Theory*, Chapman & Hall, London (1983).
14. Vasallo, C., *Optical Waveguide Cocepts*, Elsevier, Amsterdam (1991).
15. März, R., *Integrated Optics: Design and Modeling*, Artech House, Boston (1994).
16. Rozzi, T. and Mongiardo, M., *Open Electromagnetic Waveguides*, Volume 43 of IEE Electromagnetic Waves Series, The Institution of Electrical Engineers, London (1996).
17. Okamoto, K., *Fundamentals of Optical Waveguides*, Academic Press, London (2000).
18. Meade, R.D. and Winn, J.N., and Joannopoulos, J.D., *Photonic Crystals: Molding the Flow of Light*, Princeton University Press, Princeton, NJ (1995).
19. Sakoda, K., *Optical Properties of Photonic Crystals*, Volume 80 of Optical Sciences, Springer-Verlag, Berlin (2001).
20. Johnson, S.G. and Joannopoulos, J.D., Block-iterative frequency-domain methods for Maxwell's equations in planewave basis, Optics Express 8, 173–190 (2001).
21. Schmidt, F., Friese, T., Geng, R., and Zschiedrich, L., *ModeLab, a Vectorial Finite-Element Solver Based on Nodal Elements*, Konrad-Zuse Zentrum für Informationstechnik, Berlin (2000).
22. Berenger, J.P., A perfectly matched layer for the absorption of electromagnetic waves, J. Comput. Phys. 114, 185–200 (1994).
23. Taflove, A., *Computational Electrodynamics: The Finite Difference Time-Domain Method*, Artech House, Boston (1995).
24. Hess, O., Theory and simulation of spatially extended semiconductor lasers, in *Fundamental Nonlinear Laser Dynamics*, American Institute of Physics, Melville, NY (2000).
25. Nolting, H.-P. and März, R., Result of benchmark tests for different numerical BPM algorithms, IEEE J. Lightwave Technol. 13, 216–224 (1995).
26. Nolting, P., Haes, J., and Helfert, S., Benchmark tests and modelling tasks, in *Photonic Devices*, Springer-Verlag, Berlin (1999).
27. Hadley, G.R., Multistep method for wide angle beam propagation, Optics Lett. 17, 1743–1745 (1992).
28. Nedelec, J.-C., *Acoustic and Electromagnetic Equations,* Volume 144 of *Applied Mathematical Sciences*, Springer-Verlag, Berlin (2001).
29. Schmidt, F., A new approach to coupled interior-exterior Helmholtz-type problems: Theory and algorithms, Postdoctoral thesis, Freie Universität Berlin, Berlin, 2002.
30. Nishihara, H., Haruna, M., and Suhara, T., *Optical Integrated Circuits*, McGraw-Hill, New York (1989).

31. Ladouceur, F. and Love, J.D., *Silica-Based Burried Channel Waveguides and Devices*, Chapman & Hall, London (1996).
32. Madsen, C.K. and Zhao, J.H., *Optical Filter Design and Analysis*, Wiley, New York (1999).
33. März, R., Planare optische Schaltungen, in *Optische Kommunikationstechnik*, Springer-Verlag, Berlin (2002).
34. Collin, R.E., *Foundations for Microwave Engineering*, 2nd ed., IEEE Press, New York (1992).
35. März, R. and Hilliger, G., On the transfer matrix theory of contradirectional devices, Opt. Quantum Electron. 32, 829–842 (2002).
36. Cahill, L.W., The synthesis of generalized Mach-Zehnder switches based on multimode interference (MMI) couplers, Opt. Quantum Electron. 35, 465–473 (2003).
37. Garcia-Martin, A., Hermann, D., Hagmann, F., Busch, K., and Wölfle, P., Defect computations in photonic crystal: a solid state theoretical approach, Nanotechnology 14, 177–183 (2003).

Modeling of Waveguide-Optical Devices by the Method of Lines

Reinhold Pregla

FernUniversität, Fachgebiet Allgemeine und Theoretische Elektrotechnik,
Universitätstr. 27, 58084 Hagen, Germany
R.Pregla@Fernuni-Hagen.de

Planar optical waveguides are multilayered, usually even with layers of non-constant thickness. The layers may be inhomogeneous and anisotropic [1]. A typical cross section is sketched in Figure 1. Optical circuits can be understand as concatenations of waveguide sections.

Figure 2 shows an S-bend. Other typical optical waveguide circuits are, for example, Bragg gratings [2], beam splitters, and polarization converters [3]. These devices could also be concatenated to obtain more complex circuits. There are many other complex optical structures, for example, multilayered fibers (Bragg fibers), fibers with noncircular (e.g., elliptical) cross sections [4] or others (e.g., star form fibers and also fibers with an arbitrary form of the cross section [5, 6]). Very complex optical devices are vertical cavity surface emitting laser (VCSEL) diodes which consist of a very high number of different layers [7].

Now, we want to examine all of the structures by a single numerical algorithm. Let us have a look at the device in Figure 2. The optical circuits are very long compared to the transversal dimensions. The problem is to model the device with high numerical accuracy. Another important issue is the numerical stability; the device should be modeled without numerical problems. It will be demonstrated that this can be achieved with the Method of Lines (MoL) [8, 9] using the concept of impedance or admittance transformation obtained on the basis of generalized transmission line (GTL) equations [10, 11]. This transformation will be performed analytically as follows:

- From one side of a section to the other one, or from port B to port A
- From one side of a concatenation to the other side, or from 2 to 1

In summary, we transform the impedances or admittances from the output (load) to the input. Then, the field has to be calculated in the other direction, from the input toward the output. For eigenmode calculations, we perform these impedance or admittance transformations toward a common interface, where we formulate the eigenvalue problem. In the case of planar cross sections (see Fig. 1), this is done from upper and lower sides, or in case of fibers, from

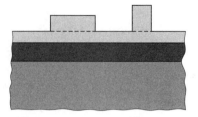

Fig. 1. Cross section of a multilayered planar optical waveguide.

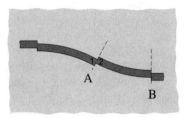

Fig. 2. S-Bend formed by concatenated waveguide sections.

outer and inner sides. These are numerically stable algorithms. To develop all of the algorithms, we start with the GTL equations [12]. For our case, the GTL equations for the transverse electric and magnetic fields must be given in matrix notation. The author has introduced these equations into modeling algorithms in 1999 [10, 11, 13] as an extension of the algorithm described in Ref. [14]. These algorithms can also successfully be used for microwave and millimeterwave devices [15–17]. Here, these equations are given in arbitrary orthogonal coordinate systems with general anisotropy of the material [18,19]. To describe the field relation between the ports of the waveguide segments, open-circuit impedance or z matrices and short-circuit admittances or y matrices are developed very easily on the basis of the GTL equations. A new algorithm will be described for periodic structures. By using Floquet modes, an impedance or admittance transformation can be performed, too, in one step from the end to the input of the periodic structure. The algorithms will be substantiated by numerical results.

1 Basic Theory: Generalized Transmission Line Equations

In this section, we derive the GTL equations which are analogous to the well-known equations for multiconductor transmission lines [20] or extended versions of them.

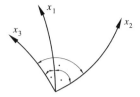

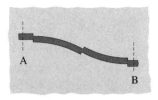

Fig. 3. General orthogonal coordinate system.

Fig. 4. Device of concatenated waveguide sections.

1.1 Material Properties

For the formulation of GTL equations the material parameter tensors can have a general form. In Eq. (1), the general permittivity/permeability tensor ($\nu = \epsilon$ or μ) and a special form are given for Cartesian coordinates:

$$\overset{\leftrightarrow}{\nu}_r = \begin{bmatrix} \nu_{xx} & \nu_{xy} & \nu_{xz} \\ \nu_{yx} & \nu_{yy} & \nu_{yz} \\ \nu_{zx} & \nu_{zy} & \nu_{zz} \end{bmatrix} \longrightarrow \overset{\leftrightarrow}{\nu}_r = \begin{bmatrix} \nu_{xx} & \nu_{xy} & 0 \\ \nu_{yx} & \nu_{yy} & 0 \\ 0 & 0 & \nu_{zz} \end{bmatrix}. \tag{1}$$

In general, orthogonal coordinates x, y and z have to be replaced by x_1, x_2 and x_3, respectively. With the special form given on the right side, gyrotropic behavior can be described. The propagation takes place, for example, in the z or x_3 direction. The device under study is divided into homogeneous sections in the direction of propagation. Hence, the material parameters in the cross sections are functions of the transversal coordinates (the coordinates in the cross section) only. In case of an eigenmode calculation, the layers are also homogeneous in the z direction.

1.2 Maxwell's Equations in Matrix Notation

For the determination of GTL equations, it is not only very efficient but also highly necessary to write the Maxwell's equations (especially curl equations) in matrix notation. This is the best way to develop the GTL equation in an analog form to the transmission line equations. First, the curl operator in general orthogonal coordinates (see Fig. 3) may be written in matrix notation in following form:

$$\overline{\nabla} \times \mathbf{A} = g_\mathrm{p}^{-1} \begin{bmatrix} 0 & -g_1 D_{\overline{x}_3} & g_1 D_{\overline{x}_2} \\ g_2 D_{\overline{x}_3} & 0 & -g_2 D_{\overline{x}_1} \\ -g_3 D_{\overline{x}_2} & g_3 D_{\overline{x}_1} & 0 \end{bmatrix} \begin{bmatrix} g_1 A_1 \\ g_2 A_2 \\ g_3 A_3 \end{bmatrix}, \tag{2}$$

where A_i and g_i ($i = 1, 2, 3$) are the field components and the metric factors, respectively. Additionally, we used the abbreviations $D_{\overline{x}_i} = \partial/\partial \overline{x}_i$ and $g_\mathrm{p} = g_1 g_2 g_3$. We have normalized the coordinates x_i with the free-space wave number k_0 (i.e., $\overline{x}_i = k_0 x_i$). Using the definitions

$$[\widehat{D}_c] = \begin{bmatrix} 0 & -D_{\overline{x}_3} & D_{\overline{x}_2} \\ D_{\overline{x}_3} & 0 & -D_{\overline{x}_1} \\ -D_{\overline{x}_2} & D_{\overline{x}_1} & 0 \end{bmatrix}, \qquad \begin{aligned} [A] &= [A_1, A_2, A_3]^t, \\ [G] &= \mathrm{diag}(g_1, g_2, g_3), \\ \end{aligned} \qquad (3)$$

$$[\widehat{D}_d] = \begin{bmatrix} D_{\overline{x}_1} & D_{\overline{x}_2} & D_{\overline{x}_3} \end{bmatrix}, \qquad \overline{\rho} = g_p \rho,$$

we obtain for Eq. (2) in shorter form,

$$\overline{\nabla} \times \mathbf{A} = g_p^{-1}[G][\widehat{D}_c][A_n], \qquad [A_n] = [G][A]. \qquad (4)$$

Maxwell's curl equations and also the divergence relations in an arbitrary orthogonal coordinate system can now be written in the following matrix form, where the magnetic field components H are normalized with the free-space wave impedance $\eta_0 = \sqrt{\mu_0/\epsilon_0}$ (i.e., $\widetilde{H} = \eta_0 H$):

$$\begin{aligned} \mathrm{j}[\overline{\epsilon}_r][E_n] &= [\widehat{D}_c][\widetilde{H}_n], \qquad & [\widehat{D}_d][\overline{\epsilon}_r][E_n] &= \overline{\rho}/\epsilon_0, \\ -\mathrm{j}[\overline{\mu}_r][\widetilde{H}_n] &= [\widehat{D}_c][E_n], \qquad & [\widehat{D}_d][\overline{\mu}_r][\widetilde{H}_n] &= 0, \end{aligned} \qquad (5)$$

with $\left(\overset{\leftrightarrow}{\nu}_r = \overset{\leftrightarrow}{\epsilon}_r, \overset{\leftrightarrow}{\mu}_r \right)$

$$[\overline{\nu}_r] = g_p\, [G]^{-1} \overset{\leftrightarrow}{\nu}_r\, [G]^{-1} = \begin{bmatrix} g_2 g_3 g_1^{-1}\nu_{11} & g_3\nu_{12} & g_2\nu_{13} \\ g_3\nu_{21} & g_1 g_3 g_2^{-1}\nu_{22} & g_1\nu_{23} \\ g_2\nu_{31} & g_1\nu_{32} & g_1 g_2 g_3^{-1}\nu_{33} \end{bmatrix} \qquad (6)$$

Equations (5) have exactly the same form as in Cartesian coordinates. Here, we have only to use normalized field components and normalized permeability and permittivity tensors. Therefore, we can directly transform the GTL equation from Cartesian coordinates [18] into an arbitrary orthogonal coordinate system. Note that $[\widehat{D}_d][\widehat{D}_c] = [0, 0, 0]$. In charge-free regions, the divergence relations are fulfilled by the curl relations.

1.3 Generalized Transmission Line Equations in Arbitrary Orthogonal Coordinate Systems and General Anisotrophy

Generalized transmission line equations can be derived for any orthogonal coordinate system [15]. Here, we will write the formulas for the case of general anisotropy [18,19]. By using the abbreviations for the transversal field vectors

$$\widehat{\mathbf{E}} = [E_{xn_1}, E_{xn_2}]^t, \qquad \widehat{\mathbf{H}} = \left[\widetilde{H}_{xn_2}, -\widetilde{H}_{xn_1} \right]^t, \qquad (7)$$

where we have ordered the components in such a way that the inner product results in the Poynting vector in the x_3 direction, we obtain from Maxwells curl equations in the case of general anisotropy for propagation (or analytical solution) in the x_3 direction:

$$\frac{\partial}{\partial \overline{x}_3}\widehat{\mathbf{E}} = -\mathrm{j}\,[R_H^{x_3}]\,\widehat{\mathbf{H}} - [S_E^{x_3}]\,\widehat{\mathbf{E}}, \qquad \frac{\partial}{\partial \overline{x}_3}\widehat{\mathbf{H}} = -\mathrm{j}\,[R_E^{x_3}]\,\widehat{\mathbf{E}} - [S_H^{x_3}]\,\widehat{\mathbf{H}}. \qquad (8)$$

On the right side of each equation, we have – compared with the multiconductor transmission lines – an additional term, the matrices $S_{E,H}$, proportional to the vector on the left side (see also Ref. [18]). These matrices $S_{E,H}$ vanish in the case of gyrotropic material tensors ($\nu_{13} = \nu_{31} = \nu_{23} = \nu_{32} = 0$) with respect to x_3 (see, e.g., Ref. [17]). We have obtained these equations from the first two scalar equations in Eq. (5), where we introduced the third one. The components E_{xn_3} and H_{xn_3} are given by

$$E_{xn_3} = -\bar{\epsilon}_{33}^{-1} \left[\bar{\epsilon}_{31}\ \bar{\epsilon}_{32} \right] \hat{E} - j\bar{\epsilon}_{33}^{-1} \left[D_{\bar{x}_1}\ D_{\bar{x}_2} \right] \hat{H}, \tag{9}$$

$$\tilde{H}_{xn_3} = -\bar{\mu}_{33}^{-1} \left[\bar{\mu}_{32}\ -\bar{\mu}_{31} \right] \hat{H} + j\bar{\mu}_{33}^{-1} \left[-D_{\bar{x}_2}\ D_{\bar{x}_1} \right] \hat{E}. \tag{10}$$

The matrices $[R_{E,H}^{x_3}]$ and $[S_{E,H}^{x_3}]$ contain derivatives with respect to x_1 and x_2 as well as the material parameters:

$$[R_H^{x_3}] = \begin{bmatrix} D_{\bar{x}_1}\bar{\epsilon}_{33}^{-1}D_{\bar{x}_1} + \bar{\mu}_{r22} & D_{\bar{x}_1}\bar{\epsilon}_{33}^{-1}D_{\bar{x}_2} - \bar{\mu}_{r21} \\ D_{\bar{x}_2}\bar{\epsilon}_{33}^{-1}D_{\bar{x}_1} - \bar{\mu}_{r12} & D_{\bar{x}_2}\bar{\epsilon}_{33}^{-1}D_{\bar{x}_2} + \bar{\mu}_{r11} \end{bmatrix} - \begin{bmatrix} \bar{\mu}_{23}\bar{\mu}_{33}^{-1}\bar{\mu}_{32} & -\bar{\mu}_{23}\bar{\mu}_{33}^{-1}\bar{\mu}_{31} \\ -\bar{\mu}_{13}\bar{\mu}_{33}^{-1}\bar{\mu}_{32} & \bar{\mu}_{13}\bar{\mu}_{33}^{-1}\bar{\mu}_{31} \end{bmatrix}, \tag{11}$$

$$[R_E^{x_3}] = \begin{bmatrix} D_{\bar{x}_2}\bar{\mu}_{33}^{-1}D_{\bar{x}_2} + \bar{\epsilon}_{r11} & \bar{\epsilon}_{r12} - D_{\bar{x}_2}\bar{\mu}_{33}^{-1}D_{\bar{x}_1} \\ \bar{\epsilon}_{r21} - D_{\bar{x}_1}\bar{\mu}_{33}^{-1}D_{\bar{x}_2} & D_{\bar{x}_1}\bar{\mu}_{33}^{-1}D_{\bar{x}_1} + \bar{\epsilon}_{r22} \end{bmatrix} - \begin{bmatrix} \bar{\epsilon}_{13}\bar{\epsilon}_{33}^{-1}\bar{\epsilon}_{31} & \bar{\epsilon}_{13}\bar{\epsilon}_{33}^{-1}\bar{\epsilon}_{32} \\ \bar{\epsilon}_{23}\bar{\epsilon}_{33}^{-1}\bar{\epsilon}_{31} & \bar{\epsilon}_{23}\bar{\epsilon}_{33}^{-1}\bar{\epsilon}_{32} \end{bmatrix}, \tag{12}$$

$$[S_H^{x_3}] = \begin{bmatrix} \bar{\epsilon}_{13} \\ \bar{\epsilon}_{23} \end{bmatrix} \left[\bar{\epsilon}_{33}^{-1}D_{\bar{x}_1}\ \bar{\epsilon}_{33}^{-1}D_{\bar{x}_2} \right] + \begin{bmatrix} D_{\bar{x}_2} \\ -D_{\bar{x}_1} \end{bmatrix} \left[\bar{\mu}_{33}^{-1}\bar{\mu}_{32}\ -\bar{\mu}_{33}^{-1}\bar{\mu}_{31} \right], \tag{13}$$

$$[S_E^{x_3}] = \begin{bmatrix} D_{\bar{x}_1} \\ D_{\bar{x}_2} \end{bmatrix} \left[\bar{\epsilon}_{33}^{-1}\bar{\epsilon}_{31}\ \bar{\epsilon}_{33}^{-1}\bar{\epsilon}_{32} \right] + \begin{bmatrix} -\bar{\mu}_{23} \\ \bar{\mu}_{13} \end{bmatrix} \left[-\bar{\mu}_{33}^{-1}D_{\bar{x}_2}\ \bar{\mu}_{33}^{-1}D_{\bar{x}_1} \right]. \tag{14}$$

For symmetrical tensor matrices $\overset{\leftrightarrow}{\mu} = \overset{\leftrightarrow}{\mu}{}^t, \overset{\leftrightarrow}{\epsilon} = \overset{\leftrightarrow}{\epsilon}{}^t$, we obtain

$$[S_E] = [S_H]^t \text{ and } [R_{E,H}] = [R_{E,H}]^t.$$

For the solution, we define a field supervector according to $\hat{F} = \left[\hat{E}, \hat{H} \right]^t$ and combine Eqs. (8):

$$\frac{\partial}{\partial \bar{x}_3}\hat{F} - \hat{Q}\hat{F} = 0, \qquad \hat{Q} = \begin{bmatrix} S_E & jR_H \\ jR_E & S_H \end{bmatrix}. \tag{15}$$

The principal solution of this equation and further steps are described in Ref. [19]. Here, we will only give the formulas for the Cartesian case and the material tensors as in Eq. (1). Using the definitions

$$\hat{E} = [E_y, E_x]^t, \qquad \hat{H} = \left[-\tilde{H}_x, \tilde{H}_y \right]^t, \tag{16}$$

where again the components are ordered in such a way that the inner product results in the Poynting vector in z direction, we obtain from Maxwell's curl equations

$$\frac{\partial}{\partial z}\widehat{H} = -j\,[R_E]\,\widehat{E}, \qquad [R_E] = \begin{bmatrix} \epsilon_{yy} + D_{\overline{x}}\mu_{zz}^{-1}D_{\overline{x}} & \epsilon_{yx} - D_{\overline{x}}\mu_{zz}^{-1}D_{\overline{y}} \\ \epsilon_{xy} - D_{\overline{y}}\mu_{zz}^{-1}D_{\overline{x}} & \epsilon_{xx} + D_{\overline{y}}\mu_{zz}^{-1}D_{\overline{y}} \end{bmatrix}, \quad (17)$$

$$\frac{\partial}{\partial z}\widehat{E} = -j\,[R_H]\,\widehat{H}, \qquad [R_H] = \begin{bmatrix} \mu_{xx} + D_{\overline{y}}\epsilon_{zz}^{-1}D_{\overline{y}} & \mu_{xy} + D_{\overline{y}}\epsilon_{zz}^{-1}D_{\overline{x}} \\ \mu_{yx} + D_{\overline{x}}\epsilon_{zz}^{-1}D_{\overline{y}} & \mu_{yy} + D_{\overline{x}}\epsilon_{zz}^{-1}D_{\overline{x}} \end{bmatrix}. \quad (18)$$

$D_{\overline{x},\overline{y}}$ are abbreviations for $\partial/\partial\overline{x},\overline{y}$. We have replaced the field components $\widetilde{H}_z$ and E_z using the remaining Maxwell's curl equations:

$$j\mu_{zz}\widetilde{H}_z = [-D_{\overline{x}}\ \ D_{\overline{y}}]\,\widehat{E}, \qquad\qquad j\epsilon_{zz}E_z = [D_{\overline{y}}\ \ D_{\overline{x}}]\,\widehat{H}. \qquad (19)$$

By combining Eqs. (17) and (18), which is very simple, we obtain wave equations

$$\frac{\partial^2}{\partial z^2}\widehat{E} - [Q^E]\,\widehat{E} = \begin{bmatrix} 0 \\ 0 \end{bmatrix}, \qquad \frac{\partial^2}{\partial z^2}\widehat{H} - [Q^H]\,\widehat{H} = \begin{bmatrix} 0 \\ 0 \end{bmatrix}, \qquad (20)$$

where $[Q^E] = -[R_H][R_E]$ and $[Q^H] = -[R_E][R_H]$. These products have to be calculated analytically because fourth-order derivatives cancel out. Therefore, numerical calculation reduces the accuracy. Both formulations in Eq. (20) are completely equivalent. There is no reason to prefer one of them. The fulfillment of boundary conditions at electric and magnetic walls is discussed in Ref. [17]. For a cylindrical coordinate system, the GTL equations are different depending on the direction of propagation or direction of analytical solution. For propagation in the z direction, for example, problems in fibers can be solved [13] or VCSEL diodes can be modeled [7]. For propagation in the ϕ direction, for example, circular waveguide bends or S-bends can be analyzed [11]. From the GTL equations in Eqs. (8), special equations can also be obtained for spherical, elliptic [4], or even general orthogonal coordinate systems [15, 19].

1.4 Discretization of the Field and Field Equations

We would like to solve the GTL equations according to the MoL [8, 9] by a discretization in the cross section (i.e., here in the x-y plane and by using analytical solution in propagation direction, in our case the z direction). The one- and two-dimensional discretization schemes are shown in Figure 5.

We use two, respectively four, different discretization line systems [17]. The components of the material parameter tensors are discretized on different places – also in the isotropic case. The effort does not increase if we analyze anisotropic structures compared to the isotropic case. E_x (E_y) is determined at the same points as H_y (H_x). Discretized quantities are represented by boldface letters. The discretized transversal field components are collected column by column in vectors. These total column vectors are marked by a hat ($\widehat{\ }$). The material parameters are combined in different diagonal matrices. For the approximation of the derivatives in the x and y direction we use operator matrices of central differences marked by $\circ$ or $\bullet$ ($\boldsymbol{D}_{\overline{x}}^{\circ;\bullet}$ and $\boldsymbol{D}_{\overline{y}}^{\circ;\bullet}$, repectively).

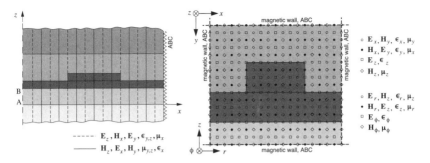

Fig. 5. Cross sections of planar optical waveguides with one- and two-dimensional discretization.

They are divided by the normalized discretization distances $\bar{h}_x = k_0 h_x$ and $\bar{h}_y = k_0 \bar{h}_y$, which are indicated by a bar over x and y. The two-dimensional discretization [marked by a hat $(\hat{\ })$] is easily described by Kronecker products. If absorbing boundary conditions (ABC) [9] are necessary, they have to be introduced instead of magnetic walls. To be more flexible with respect to the cross section, nonequidistant discretization can also be used following the hints given in Ref. [22]. With the supervectors of the discretized field components

$$\widehat{\mathbf{E}} = \left[\widehat{\mathbf{E}}_y^t, \widehat{\mathbf{E}}_x^t\right]^t, \qquad \widehat{\mathbf{H}} = \left[-\widehat{\mathbf{H}}_x^t, \widehat{\mathbf{H}}_y^t\right]^t, \tag{21}$$

the discretized Eqs. (17) and (18) take the form – where we assumed only diagonal material tensors –

$$\frac{\mathrm{d}}{\mathrm{d}\bar{z}}\widehat{\mathbf{H}} = -\mathrm{j}\widehat{\boldsymbol{R}}_\mathrm{E}\,\widehat{\mathbf{E}}, \qquad \widehat{\boldsymbol{R}}_\mathrm{E} = \begin{bmatrix} \widehat{\boldsymbol{\epsilon}}_y - \widehat{\boldsymbol{D}}_{\bar{x}}^{\bullet t}\widehat{\boldsymbol{\mu}}_z^{-1}\widehat{\boldsymbol{D}}_{\bar{x}}^{\bullet} & \widehat{\boldsymbol{D}}_{\bar{x}}^{\bullet t}\widehat{\boldsymbol{\mu}}_z^{-1}\widehat{\boldsymbol{D}}_{\bar{y}}^{\circ} \\ \widehat{\boldsymbol{D}}_{\bar{y}}^{\mathrm{ot}}\widehat{\boldsymbol{\mu}}_z^{-1}\widehat{\boldsymbol{D}}_{\bar{x}}^{\bullet} & \widehat{\boldsymbol{\epsilon}}_x - \widehat{\boldsymbol{D}}_{\bar{y}}^{\mathrm{ot}}\widehat{\boldsymbol{\mu}}_z^{-1}\widehat{\boldsymbol{D}}_{\bar{y}}^{\circ} \end{bmatrix}, \tag{22}$$

$$\frac{\mathrm{d}}{\mathrm{d}\bar{z}}\widehat{\mathbf{E}} = -\mathrm{j}\widehat{\boldsymbol{R}}_\mathrm{H}\,\widehat{\mathbf{H}}, \qquad \widehat{\boldsymbol{R}}_\mathrm{H} = \begin{bmatrix} \widehat{\boldsymbol{\mu}}_x - \widehat{\boldsymbol{D}}_{\bar{y}}^{\bullet t}\widehat{\boldsymbol{\epsilon}}_z^{-1}\widehat{\boldsymbol{D}}_{\bar{y}}^{\bullet} & -\widehat{\boldsymbol{D}}_{\bar{y}}^{\bullet t}\widehat{\boldsymbol{\epsilon}}_z^{-1}\widehat{\boldsymbol{D}}_{\bar{x}}^{\circ} \\ -\widehat{\boldsymbol{D}}_{\bar{x}}^{\mathrm{ot}}\widehat{\boldsymbol{\epsilon}}_z^{-1}\widehat{\boldsymbol{D}}_{\bar{y}}^{\bullet} & \widehat{\boldsymbol{\mu}}_y - \widehat{\boldsymbol{D}}_{\bar{x}}^{\mathrm{ot}}\widehat{\boldsymbol{\epsilon}}_z^{-1}\widehat{\boldsymbol{D}}_{\bar{x}}^{\circ} \end{bmatrix}. \tag{23}$$

The correct order of all of the quantities is important to fulfill the interface conditions. In the case of gyrotropic (gyromagnetic or gyroelectric) materials, the terms with ϵ_{xy}, ϵ_{yx}, μ_{xy} and μ_{yx} must also be introduced. Absorbing boundary conditions [9] can be realized in these equations by suitable replacements [17]. Combining Eqs. (22) and (23), we obtain for the discretized wave equations,

$$\frac{\mathrm{d}^2}{\mathrm{d}\bar{z}^2}\widehat{\mathbf{H}} - \widehat{\boldsymbol{Q}}^\mathrm{H}\widehat{\mathbf{H}} = 0, \qquad \frac{\mathrm{d}^2}{\mathrm{d}\bar{z}^2}\widehat{\mathbf{E}} - \widehat{\boldsymbol{Q}}^\mathrm{E}\widehat{\mathbf{E}} = 0, \qquad \widehat{\boldsymbol{Q}}^\mathrm{H} = -\widehat{\boldsymbol{R}}_\mathrm{E}\,\widehat{\boldsymbol{R}}_\mathrm{H}, \qquad \widehat{\boldsymbol{Q}}^\mathrm{E} = -\widehat{\boldsymbol{R}}_\mathrm{H}\,\widehat{\boldsymbol{R}}_\mathrm{E}, \tag{24}$$

where, by calculating the submatrices, we have to take into account that $-\boldsymbol{D}_{\bar{y}}^{\bullet t} = \boldsymbol{D}_{\bar{y}}^{\circ}$ and $-\boldsymbol{D}_{\bar{x}}^{\mathrm{ot}} = \boldsymbol{D}_{\bar{x}}^{\bullet}$.

1.5 Eigenvalue and Modal Matrices

Transforming the fields by using the modal matrices $\widehat{T}_{\mathrm{H}}$ and $\widehat{T}_{\mathrm{E}}$ according to

$$\widehat{\mathbf{H}} = \widehat{T}_{\mathrm{H}}\widehat{\widehat{\mathbf{H}}}, \qquad \widehat{\mathbf{E}} = \widehat{T}_{\mathrm{E}}\widehat{\widehat{\mathbf{E}}}, \qquad \widehat{T}_{\mathrm{H}}^{-1}\widehat{\mathbf{Q}}_{\mathrm{H}}\widehat{T}_{\mathrm{H}}=\widehat{\boldsymbol{\varGamma}}^2, \qquad \widehat{T}_{\mathrm{E}}^{-1}\widehat{\mathbf{Q}}_{\mathrm{E}}\widehat{T}_{\mathrm{E}}=\widehat{\boldsymbol{\varGamma}}^2, \qquad (25)$$

Eqs. (24) reduce to

$$\frac{\mathrm{d}^2}{\mathrm{d}\bar{z}^2}\widehat{\widehat{\mathbf{H}}} - \widehat{\boldsymbol{\varGamma}}^2\widehat{\widehat{\mathbf{H}}} = \mathbf{0}, \qquad \frac{\mathrm{d}^2}{\mathrm{d}\bar{z}^2}\widehat{\widehat{\mathbf{E}}} - \widehat{\boldsymbol{\varGamma}}^2\widehat{\widehat{\mathbf{E}}} = \mathbf{0}. \qquad (26)$$

We obtain from both eigenvalue problems in Eqs. (25) identical diagonal matrices of propagation constants $\boldsymbol{\varGamma}^2$. This relations holds because both Q matrices are obtained as products of the R matrices but in different order. The general relation between the eigenvector matrices $\widehat{T}_{\mathrm{E}}$ and $\widehat{T}_{\mathrm{H}}$ of $\boldsymbol{Q}_{\mathrm{E}}$ and $\boldsymbol{Q}_{\mathrm{H}}$, respectively, is

$$\boldsymbol{T}_{\mathrm{E}} = \widehat{\mathbf{R}}_{\mathrm{H}}\boldsymbol{T}_{\mathrm{H}}\widehat{\beta}_1^{-1}, \qquad \boldsymbol{T}_{\mathrm{H}} = \widehat{\mathbf{R}}_{\mathrm{E}}\boldsymbol{T}_{\mathrm{E}}\widehat{\beta}_2^{-1}, \quad \widehat{\beta}_1\widehat{\beta}_2 = -\widehat{\boldsymbol{\varGamma}}^2, \quad \widehat{\beta}_1 = \widehat{\beta}_2 = \widehat{\beta}, \qquad (27)$$

which is known from the theory of matrices. Generally, the amplitudes of the eigenvectors are free; therefore, we have introduced the (diagonal) matrices $\widehat{\beta}_{1,2}^{-1}$ for normalization. The diagonal matrices $\widehat{\beta}_1$ and $\widehat{\beta}_2$ must fulfill the above condition. By transforming the GTL equations (22) and (23) with the two modal matrices, we obtain simple transformed GTL equations

$$\frac{\mathrm{d}}{\mathrm{d}\bar{z}}\widehat{\widehat{\mathbf{E}}} = -\mathrm{j}\widehat{\beta}\,\widehat{\widehat{\mathbf{H}}}, \qquad \frac{\mathrm{d}}{\mathrm{d}\bar{z}}\widehat{\widehat{\mathbf{H}}} = -\mathrm{j}\widehat{\beta}\,\widehat{\widehat{\mathbf{E}}}. \qquad (28)$$

2 Impedance/Admittance Transformation

2.1 Transformation Through Waveguide Sections

From the general solution of Eqs. (26), $\widehat{\widehat{\mathbf{F}}} = \widehat{\widehat{\mathbf{F}}}_{\mathrm{f}} + \widehat{\widehat{\mathbf{F}}}_{\mathrm{b}} = \exp(-\widehat{\boldsymbol{\varGamma}}\bar{z})\mathbf{A} + \exp(\widehat{\boldsymbol{\varGamma}}\bar{z})\mathbf{B}$, where $\widehat{\widehat{\mathbf{F}}} = \widehat{\widehat{\mathbf{E}}},\widehat{\widehat{\mathbf{H}}}$, we obtain a relation between the derivatives of the fields and the fields themselves in the two cross sections A and B (inner sides) of a longitudinal homogeneous section whose distance is d $(\bar{d} = k_0 d)$:

$$\frac{\mathrm{d}}{\mathrm{d}\bar{z}}\begin{bmatrix}\widehat{\widehat{\mathbf{F}}}_{\mathrm{A}} \\ \widehat{\widehat{\mathbf{F}}}_{\mathrm{B}}\end{bmatrix} = \begin{bmatrix}-\widehat{\gamma} & \widehat{\alpha} \\ -\widehat{\alpha} & \widehat{\gamma}\end{bmatrix}\begin{bmatrix}\widehat{\widehat{\mathbf{F}}}_{\mathrm{A}} \\ \widehat{\widehat{\mathbf{F}}}_{\mathrm{B}}\end{bmatrix}, \qquad \begin{aligned}\widehat{\alpha}&=\widehat{\boldsymbol{\varGamma}}/\sinh(\widehat{\boldsymbol{\varGamma}}\bar{d}), \\ \widehat{\gamma}&=\widehat{\boldsymbol{\varGamma}}/\tanh(\widehat{\boldsymbol{\varGamma}}\bar{d}).\end{aligned} \qquad (29)$$

Using the first part of the general solution (the forward propagating fields $\widehat{\widehat{\mathbf{F}}}_{\mathrm{f}}$), we can define wave impedance/admittance matrices: $\widehat{\overline{\mathbf{Z}}}_0 = \boldsymbol{I}$ and $\widehat{\overline{\mathbf{Y}}}_0 = \boldsymbol{I}$. This simple result is a consequence of the field normalization using the two modal matrices $\boldsymbol{T}_{\mathrm{E}}$ and $\boldsymbol{T}_{\mathrm{H}}$. Introducing Eqs. (28) into (29) results in

$$\begin{bmatrix} \widehat{\overline{\mathbf{H}}}_A \\ -\widehat{\overline{\mathbf{H}}}_B \end{bmatrix} = \begin{bmatrix} \overline{\mathbf{y}}_1 & \overline{\mathbf{y}}_2 \\ \overline{\mathbf{y}}_2 & \overline{\mathbf{y}}_1 \end{bmatrix} \begin{bmatrix} \widehat{\overline{\mathbf{E}}}_A \\ \widehat{\overline{\mathbf{E}}}_B \end{bmatrix}, \qquad \begin{bmatrix} \widehat{\overline{\mathbf{E}}}_A \\ \widehat{\overline{\mathbf{E}}}_B \end{bmatrix} = \begin{bmatrix} \widehat{\mathbf{z}}_1 & \widehat{\mathbf{z}}_2 \\ \widehat{\mathbf{z}}_2 & \widehat{\mathbf{z}}_1 \end{bmatrix} \begin{bmatrix} \widehat{\overline{\mathbf{H}}}_A \\ -\widehat{\overline{\mathbf{H}}}_B \end{bmatrix}, \qquad (30)$$

where

$$\begin{aligned} \overline{\mathbf{y}}_1 &= \widehat{\overline{\mathbf{Y}}}_0/\tanh(\widehat{\overline{\boldsymbol{\Gamma}}}\overline{d}), & \overline{\mathbf{y}}_2 &= -\widehat{\overline{\mathbf{Y}}}_0/\sinh(\widehat{\overline{\boldsymbol{\Gamma}}}\overline{d}), \\ \widehat{\mathbf{z}}_1 &= \widehat{\overline{\mathbf{Z}}}_0/\tanh(\widehat{\overline{\boldsymbol{\Gamma}}}\overline{d}), & \widehat{\mathbf{z}}_2 &= \widehat{\overline{\mathbf{Z}}}_0/\sinh(\widehat{\overline{\boldsymbol{\Gamma}}}\overline{d}). \end{aligned} \qquad (31)$$

We obtain open-circuit and short-circuit parameters as known from the circuit theory. Here, these parameters are diagonal matrices. Defining admittances/impedances according to

$$\widehat{\overline{\mathbf{H}}}_{A,B} = \widehat{\overline{\mathbf{Y}}}_{A,B}\,\widehat{\overline{\mathbf{E}}}_{A,B}, \qquad \widehat{\overline{\mathbf{E}}}_{A,B} = \widehat{\overline{\mathbf{Z}}}_{A,B}\,\widehat{\overline{\mathbf{H}}}_{A,B} \qquad (32)$$

results in the admittance/impedance transformation between cross sections A and B:

$$\widehat{\overline{\mathbf{Y}}}_A = \widehat{\overline{\mathbf{y}}}_1 - \widehat{\overline{\mathbf{y}}}_2 \left(\widehat{\overline{\mathbf{y}}}_1 + \widehat{\overline{\mathbf{Y}}}_B\right)^{-1}\widehat{\overline{\mathbf{y}}}_2, \qquad \widehat{\overline{\mathbf{Z}}}_A = \widehat{\mathbf{z}}_1 - \widehat{\mathbf{z}}_2 \left(\widehat{\mathbf{z}}_1 + \widehat{\overline{\mathbf{Z}}}_B\right)^{-1}\widehat{\mathbf{z}}_2. \qquad (33)$$

These admittance/impedance transformation formulas are numerically stable.

2.2 Waveguide Discontinuities

If waveguide sections with different cross sections are concatenated, the tangential fields in the common cross section have to be matched. The matching process must be performed in the original domain. The related impedance transformation is given by [17]

$$\widehat{\mathbf{Z}}_1 = \mathbf{T}_{\mathrm{E1}}^{-1}\mathbf{T}_{\mathrm{E2}}\widehat{\mathbf{Z}}_2\mathbf{T}_{\mathrm{H2}}^{-1}\mathbf{T}_{\mathrm{H1}}. \qquad (34)$$

2.3 Concatenations of Waveguide Sections

Concatenations of waveguide sections (see Fig. 4) can also be described by $\mathbf{y}$ and $\mathbf{z}$ matrices. Again, they have the same form as in transmission line or in circuit theory. Therefore, the submatrices $\mathbf{z}_{ik}$ are also obtained as in circuit theory. By choosing $\mathbf{H}_B$ ($\mathbf{H}_A$) equal zero (magnetic wall at port B (A)), the submatrices $\widehat{\mathbf{z}}_{11}$ and $\widehat{\mathbf{z}}_{12}$ ($\widehat{\mathbf{z}}_{21}$ and $\widehat{\mathbf{z}}_{22}$) can be determined. In the case of $\mathbf{y}$ matrices, the analogous short-circuit 'experiments' have to be performed. Concatenations of sections can also be described by using the chain matrix description, which can be obtained from the above impedance/admittance matrices. We write here these relation in the original domain:

$$\begin{bmatrix} \widehat{\mathbf{E}}_A \\ \widehat{\mathbf{E}}_B \end{bmatrix} = \begin{bmatrix} \widehat{\mathbf{z}}_{11} & \widehat{\mathbf{z}}_{12} \\ \widehat{\mathbf{z}}_{21} & \widehat{\mathbf{z}}_{22} \end{bmatrix} \begin{bmatrix} \widehat{\mathbf{H}}_A \\ -\widehat{\mathbf{H}}_B \end{bmatrix}, \qquad \begin{bmatrix} \widehat{\mathbf{E}}_B \\ \widehat{\mathbf{H}}_B \end{bmatrix} = \begin{bmatrix} \widehat{\mathbf{z}}_{22}\widehat{\mathbf{z}}_{12}^{-1} & \widehat{\mathbf{z}}_{21} - \widehat{\mathbf{z}}_{22}\widehat{\mathbf{z}}_{12}^{-1}\widehat{\mathbf{z}}_{11} \\ -\widehat{\mathbf{z}}_{12}^{-1} & \widehat{\mathbf{z}}_{12}^{-1}\widehat{\mathbf{z}}_{11} \end{bmatrix} \begin{bmatrix} \widehat{\mathbf{E}}_A \\ \widehat{\mathbf{H}}_A \end{bmatrix}. \qquad (35)$$

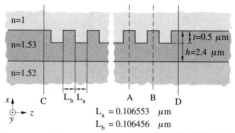

Fig. 6. Bragg grating as analyzed in the COST 240 program.

In the case of a symmetric two-port between A and B, the two matrices $\widehat{\mathbf{z}}_{11}$ and $\widehat{\mathbf{z}}_{12}$ can be obtained from $\widehat{\mathbf{z}}_{11h}$ and $\widehat{\mathbf{y}}_{11h}$, the matrix parameters of the half two-port. The subscript h symbolizes the half of the two-port.

$$\widehat{\mathbf{z}}_{11} = \tfrac{1}{2}\left(\widehat{\mathbf{z}}_{11h} + \widehat{\mathbf{y}}_{11h}^{-1}\right), \qquad \widehat{\mathbf{z}}_{12} = \tfrac{1}{2}\left(\widehat{\mathbf{z}}_{11h} - \widehat{\mathbf{y}}_{11h}^{-1}\right). \tag{36}$$

The fields in the structure can be computed accurately and without numerical problems [14]. The algorithm is an analog to the one used in transmission line problems. There, we transform the impedances from the end of the line to the input. Then, we can transform the currents and voltages in the opposite direction from the input to the output.

3 Periodic Structures

3.1 Floquet Modes of a Period

Periodic structures in optical devices often consist of a very large number of periods. In Figure 6, a Bragg grating is shown as an example. In these cases, it makes no sense to transform the impedances/admittances period by period. A better way is to determine the Floquet modes of a single period. Then, the total periodic structure can be described as a "homogeneous" waveguide. In this description, the period should be symmetric with $\widehat{\mathbf{z}}_{22} = \widehat{\mathbf{z}}_{11}$ and $\widehat{\mathbf{z}}_{21} = \widehat{\mathbf{z}}_{12}$. We perform a transformation to Floquet modes according to

$$\widehat{\mathbf{E}}_{A,B} = \mathbf{S}_E \widetilde{\mathbf{E}}_{A,B}, \qquad \widehat{\mathbf{H}}_{A,B} = \mathbf{S}_H \widetilde{\mathbf{H}}_{A,B}. \tag{37}$$

Equations (35) should be equivalent to the equation of Floquet modes

$$\begin{bmatrix} \widetilde{\mathbf{E}}_B \\ \widetilde{\mathbf{H}}_B \end{bmatrix} = \begin{bmatrix} \cosh \boldsymbol{\Gamma}_F & -\widetilde{\mathbf{Z}}_0 \sinh \boldsymbol{\Gamma}_F \\ -\widetilde{\mathbf{Y}}_0 \sinh \boldsymbol{\Gamma}_F & \cosh \boldsymbol{\Gamma}_F \end{bmatrix} \begin{bmatrix} \widetilde{\mathbf{E}}_A \\ \widetilde{\mathbf{H}}_A \end{bmatrix} \tag{38}$$

To obtain this equivalence, we have to solve the following eigenvalue problems:

$$-\widehat{\mathbf{y}}_{12}^{-1}\widehat{\mathbf{y}}_{11}\mathbf{S}_E = \widehat{\mathbf{z}}_{11}\widehat{\mathbf{z}}_{12}^{-1}\mathbf{S}_E = \mathbf{S}_E \boldsymbol{\lambda}_E, \qquad -\widehat{\mathbf{y}}_{11}\widehat{\mathbf{y}}_{12}^{-1}\mathbf{S}_H = \widehat{\mathbf{z}}_{12}^{-1}\widehat{\mathbf{z}}_{11}\mathbf{S}_H = \mathbf{S}_H \boldsymbol{\lambda}_H, \tag{39}$$

which result in

$$\lambda_E = \lambda_H = \lambda = \cosh \Gamma_F. \tag{40}$$

S_E and S_H are not independent of each other. The following relations are valid:

$$S_E = \hat{z}_{11} S_H \delta_1, \qquad S_H = \hat{z}_{12}^{-1} S_E \delta_2, \tag{41}$$

where $\delta_{1,2}$ are diagonal matrices. For self-consistency, they have to fulfill the condition

$$\delta_1 \delta_2 = \lambda^{-1}. \tag{42}$$

By choosing $\tilde{Z}_0 = \tilde{Y}_0 = I$ (other choices, e.g., $\delta_1 = \delta_2$ are also possible), we obtain

$$\delta_1 = \sqrt{\lambda^2 - I}/\lambda, \qquad \delta_2 = I/\sqrt{\lambda^2 - I}. \tag{43}$$

Γ_F should be alternatively to Eq. (40) determined by [23]

$$\tanh \Gamma_F = I/\sqrt{S_E^{-1}\hat{z}_{11}\hat{y}_{11}S_E} \quad \text{or} \quad \tanh\left(\frac{1}{2}\Gamma_F\right) = I/\sqrt{S_E^{-1}\hat{z}_{11h}\hat{y}_{11h}S_E}. \tag{44}$$

3.2 Concatenation of N Periods

In this subsection, we give the relevant formulas for N concatenated period sections (see Fig. 6). Because we have described each section as a homogeneous waveguide for concatenation of N periods only Γ_F should be replaced by $N\Gamma_F$. Equation (38) now reads

$$\begin{bmatrix} \tilde{E}_D \\ \tilde{H}_D \end{bmatrix} = \begin{bmatrix} \cosh(N\Gamma_F) & -\tilde{Z}_0 \sinh(N\Gamma_F) \\ -\tilde{Y}_0 \sinh(N\Gamma_F) & \cosh(N\Gamma_F) \end{bmatrix} \begin{bmatrix} \tilde{E}_C \\ \tilde{H}_C \end{bmatrix}. \tag{45}$$

The ports C and D are the input and output ports of the whole structure. Furthermore, we can again write the field relations between the generalized two-ports C and D with open-circuit impedance or short-circuit admittance matrix-parameter description (z and y matrices):

$$\begin{bmatrix} \tilde{E}_C \\ \tilde{E}_D \end{bmatrix} = \begin{bmatrix} \tilde{z}_1 & \tilde{z}_2 \\ \tilde{z}_2 & \tilde{z}_1 \end{bmatrix} \begin{bmatrix} \tilde{H}_C \\ -\tilde{H}_D \end{bmatrix}, \qquad \begin{bmatrix} \tilde{H}_C \\ -\tilde{H}_D \end{bmatrix} = \begin{bmatrix} \tilde{y}_1 & \tilde{y}_2 \\ \tilde{y}_2 & \tilde{y}_1 \end{bmatrix} \begin{bmatrix} \tilde{E}_C \\ \tilde{E}_D \end{bmatrix}, \tag{46}$$

with

$$\begin{aligned} \tilde{z}_1 &= \tilde{Z}_0/\tanh(N\tilde{\Gamma}_F), & \tilde{z}_2 &= \tilde{Z}_0/\sinh(N\tilde{\Gamma}_F), \\ \tilde{y}_1 &= \tilde{Y}_0/\tanh(N\tilde{\Gamma}_F), & \tilde{y}_2 &= -\tilde{Y}_0/\sinh(N\tilde{\Gamma}_F). \end{aligned} \tag{47}$$

With $\tilde{E}_{C,D} = \tilde{Z}_{C,D} \tilde{H}_{C,D}$ and $\tilde{H}_{C,D} = \tilde{Y}_{C,D} \tilde{E}_{C,D}$, the impedance/admittance transformation for the whole periodic structure is performed by

$$\widetilde{\boldsymbol{Z}}_C = \widetilde{\boldsymbol{z}}_1 - \widetilde{\boldsymbol{z}}_2 \left(\widetilde{\boldsymbol{z}}_1 + \widetilde{\boldsymbol{Z}}_D\right)^{-1} \widetilde{\boldsymbol{z}}_2, \qquad \widetilde{\boldsymbol{Y}}_C = \widetilde{\boldsymbol{y}}_1 - \widetilde{\boldsymbol{y}}_2 \left(\widetilde{\boldsymbol{y}}_1 + \widetilde{\boldsymbol{Y}}_D\right)^{-1} \widetilde{\boldsymbol{y}}_2. \qquad (48)$$

At the output and input, the relations between the impedances/admittances with the tilde ($\sim$) (Floquet impedances) and bar ($-$) (mode impedances) are given by

$$\widetilde{\boldsymbol{Z}}_D = \boldsymbol{S}_E^{-1} \boldsymbol{T}_E \overline{\boldsymbol{Z}}_D \boldsymbol{T}_H^{-1} \boldsymbol{S}_H, \qquad \widetilde{\boldsymbol{Y}}_D = \boldsymbol{S}_H^{-1} \boldsymbol{T}_H \overline{\boldsymbol{Y}}_D \boldsymbol{T}_E^{-1} \boldsymbol{S}_E, \qquad (49)$$

which can also be easily can be inverted to yield the mode impedances/admittances from the Floquet values. The scattering parameters are obtained as in Section 3.3.

3.3 Scattering Parameters

In this subsection, we will describe the procedure to obtain the scattering parameters for the periodic structures. The general principle can also be used for other devices of concatenated waveguide sections.

By using the impedance transformation formula developed above, we can calculate the input impedance of a concatenation structure starting at the end. The load impedance matrix $\overline{\boldsymbol{Z}}_D$ in most cases is the matrix of characteristic impedances $\overline{\boldsymbol{Z}}_{0D}$ of the outgoing waveguide. If the outgoing waveguide is not of the same kind as the last section of the period, then a transformation according to Eq. (34) must be performed. At the input of the structure, we have a feeding waveguide. We assume that this waveguide has a characteristic impedance matrix $\overline{\boldsymbol{Z}}_0$. The load impedance of the input waveguide (the input impedance matrix of the structure) should be given by $\overline{\boldsymbol{Z}}_C$. For the magnetic field at the input, we may then write

$$\overline{\boldsymbol{H}}_C = 2\left(\overline{\boldsymbol{Z}}_C + \overline{\boldsymbol{Z}}_0\right)^{-1} \overline{\boldsymbol{E}}_{Af}, \qquad (50)$$

where $\overline{\boldsymbol{E}}_{Cf}$ is the vector of the propagating modes in the forward direction. If we assume that in the input section only the fundamental mode with amplitude 1 is propagating in the forward direction, then $\overline{\boldsymbol{E}}_{Cf}$ is given by

$$\overline{\boldsymbol{E}}_{Cf} = [1, 0, \dots, 0]^t. \qquad (51)$$

(The eigenmodes are ordered in such a way that the fundamental mode will be found in the first column of $\overline{\boldsymbol{T}}$.) With Eq. (50) and $\overline{\boldsymbol{E}}_C = \overline{\boldsymbol{Z}}_C \overline{\boldsymbol{H}}_C$, we obtain

$$\overline{\boldsymbol{E}}_C = 2\overline{\boldsymbol{Z}}_C (\overline{\boldsymbol{Z}}_C + \overline{\boldsymbol{Z}}_0)^{-1} \overline{\boldsymbol{E}}_{Cf}. \qquad (52)$$

The vector $\overline{\boldsymbol{E}}_C$ contains the complex amplitude of the reflected fundamental and all higher modes. The scattering coefficient column vector $\boldsymbol{S}_{11}$ is now given by

$$\boldsymbol{S}_{11} = \overline{\boldsymbol{E}}_C - \overline{\boldsymbol{E}}_{Cf}. \qquad (53)$$

By repeating this calculation for the other modes in the input waveguide, we can construct the generalized scattering matrix S_{11}. The reflection coefficient of the fundamental mode is given in the vector S_{11} as the first component. From the field at the input of the structure, we can calculate the field in the whole structure using the derived impedance/admittance transfer equations and the forward and backward propagating field parts. The algorithm is stable and highly accurate. From the fields at the end, all of the other scattering parameters can be obtained. The fields $\overline{\mathbf{E}}_C$ and $\overline{\mathbf{H}}_C$ now have to be transformed into the Floquet domain by

$$\widetilde{\mathbf{E}}_C = S_E^{-1} T_E \overline{\mathbf{E}}_C, \qquad \widetilde{\mathbf{H}}_C = S_H^{-1} T_H \overline{\mathbf{H}}_C. \tag{54}$$

Because the Floquet modes behave like normal modes, we now split the fields into forward (subscript f) and backward (subscript b) propagating parts. At port C, we obtain

$$\widetilde{\mathbf{E}}_{Cf} = \frac{1}{2} \left(\widetilde{\mathbf{E}}_C + \widetilde{\mathbf{Z}}_0 \widetilde{\mathbf{H}}_C \right), \qquad \widetilde{\mathbf{E}}_{Cb} = \frac{1}{2} \left(\widetilde{\mathbf{E}}_C - \widetilde{\mathbf{Z}}_0 \widetilde{\mathbf{H}}_C \right). \tag{55}$$

For the forward propagating part, we obtain at port D,

$$\widetilde{\mathbf{E}}_{Df} = \exp(-N\boldsymbol{\Gamma}_F)\widetilde{\mathbf{E}}_{Cf}. \tag{56}$$

It does not make sense to calculate the backward propagating part $\widetilde{\mathbf{E}}_{Db}$ from $\widetilde{\mathbf{E}}_{Cb}$ because of a positive sign in the exponent. By knowing the load impedance $\widetilde{\mathbf{Z}}_D$, we can calculate the fields at D from $\mathbf{E}_{Df}$ by

$$\widetilde{\mathbf{E}}_D = 2\widetilde{\mathbf{Z}}_D \left(\widetilde{\mathbf{Z}}_0 + \widetilde{\mathbf{Z}}_D \right)^{-1} \widetilde{\mathbf{E}}_{Df}, \qquad \widetilde{\mathbf{H}}_D = \widetilde{\mathbf{Y}}_D \widetilde{\mathbf{E}}_D. \tag{57}$$

$\widetilde{\mathbf{E}}_D$ and $\widetilde{\mathbf{H}}_D$ have to be transformed into $\overline{\mathbf{E}}_D$ and $\overline{\mathbf{H}}_D$ by

$$\overline{\mathbf{E}}_D = T_E^{-1} S_E \widetilde{\mathbf{E}}_D, \qquad \overline{\mathbf{E}}_D = T_H^{-1} S_H \widetilde{\mathbf{H}}_D. \tag{58}$$

With the condition described above, we obtain

$$S_{21} = \overline{\mathbf{E}}_D \tag{59}$$

and especially

$$S_{21} = \overline{\mathbf{E}}_D(1,1). \tag{60}$$

The algorithm described here for a periodic section can be used analogously for other waveguide sections. The Floquet quantities then have to be replaced by the quantities of the relevant waveguide section.

4 Numerical Results

Numerical results computed for special devices are given here to show the accuracy and the wide applicability of the proposed algorithm.

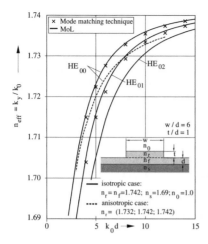

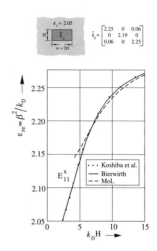

Fig. 7. Dispersion diagrams for a rib guide (From Ref. [3] © 2003 Kluwer.)

Fig. 8. Dispersion diagrams for a channel guide (From Ref. [19] © 2002 IEEE.)

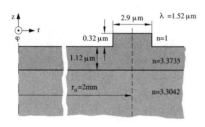

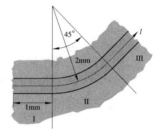

Fig. 9. Rib waveguide bend: (a) cross section; (b) top view.

Figure 7 shows a dispersion diagram for a rib guide. The effective refractive index – which is equal to the propagation constant normalized with k_0 – is drawn as function of k_0 times d or the normalized frequency for three HE modes. The results are in good agreement with those obtained with mode-matching technique. The waveguide material was isotropic in this case. The dashed curve is for anisotropic material and the HE_{00} mode.

In the next example (see Fig. 8), the effective permittivity ε_{re} for a channel guide is drawn as function of k_0 times H (normalized frequency). The channel material is anisotropic with the permittivity tensor given in Figure 8. Again, we have good agreement with other results where the structure is completely surrounded by metallic walls.

Next, results are presented for curved waveguides or concatenations of curved waveguides and straight sections. In Figure 9, the cross section and the longitudinal section of the analyzed device are given. The fundamental mode was injected and determined the field distribution. The power was then

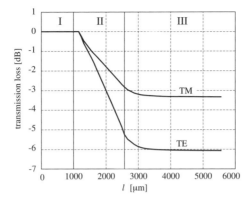

Fig. 10. Transmission of the power along the waveguide bend. TE polarization: 11.8 dB/90° loss; TM polarization: 6.7 dB/90° loss.

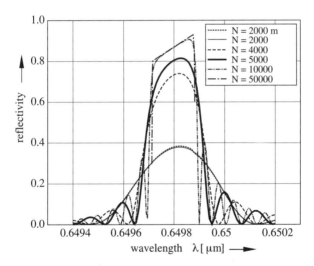

Fig. 11. Reflectivity of the Bragg grating in Figure 6 with different number of periods as function of the wavelength. (From Ref. [23] © 2003 Urban & Fischer.)

calculated by an integration of the Poynting vector in the computation window. The upper curve in Figure 10 shows results for the TM case, and on the lower curve, results for TE polarization is given. We see no losses in the input part. In the curved part, the loss curve is linear after higher modes have left the computational window. In the output, we see again that the bend modes are not matched with the output mode. Therefore, we have also some losses here until only the fundamental mode remains. The losses for the TE case are higher than those for the TM polarization.

Results for the spectral behavior of the reflectivity ($|S_{11}|^2$) for the Bragg grating in Figure 6 – analyzed with the algorithm described – are shown in

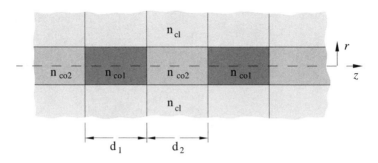

Fig. 12. Fiber grating structure: $r_{co1} = r_{co2} = 2\,\mu m$.

Figure 11 [23] for different numbers of periods N. From the coupled-mode theory, a reflectivity of more than 90 % for 1000 periods was predicted at the wavelength of 650 nm. As can be seen, a maximum of more than 90 % can be reached. However, the required number of periods is about 10,000 and the wavelength is lower than 650 nm. It should be mentioned also that such a high number of periods did not lead to any numerical problems.

It is an interesting phenomenon that the curves are not symmetrical. The results are identical to those published in Refs. [2] and [21]. The curves are calculated with absorbing boundary conditions. Also, electric and magnetic walls can be used below and above the structure. In these cases, the curves are superimposed by a very small oscillation, as can be seen for the case $N = 2000\,m$ (metallic walls). The accuracy of the calculations can especially be seen on the left and right sides of main part, which shows a smooth behavior. (This result is confirmed by the calculation for the case $N = 50000$ not drawn in Figure 11.)

The algorithm was also used to analyze waveguides in other than Cartesian coordinates. Figure 12 shows a longitudinal section of a fiber Bragg grating.

Figure 13 shows the behavior of the fiber-Bragg grating. Curves for 5000 and 8000 periods were calculated. Also, anisotropic material was introduced. The tensor component in the propagation direction is different than the transverse ones. The curve with the anisotropic materials is only shifted slightly to the right.

As example of our modeling of active devices results for VCSEL diodes [7] will be shown. VCSEL diodes are very complex structures (see Fig. 14). The main parts are the resonator, the electronic supply system, and the active region. The electronic system provides the active region with electrons and holes. The recombination process results in photons. The generation number is proportional to the square of the electric field. Let us first study the principle of the resonator. Figure 15 shows a model of a laser resonator. It consists of two Bragg mirrors. Each layer must have a thickness of $\lambda/4$. In the middle, we have a layer for phase adjustment. The thickness of this layer must be $n\lambda/2$, with $n = 0, 1, 2, \ldots$

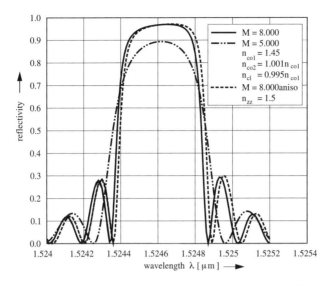

Fig. 13. Fiber grating: Reflectivity as function of wavelength. (From Ref. [19] © 2002 IEEE.)

The field in this structure is given by the components E_x and H_y. The transmission line and wave equations are obtained from Eqs. (17)–(20) and

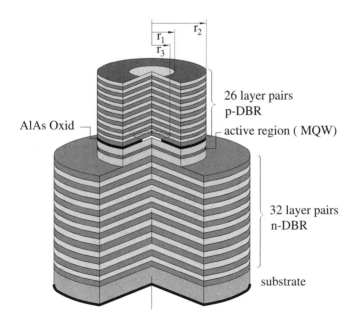

Fig. 14. Model of a VCSEL diode.

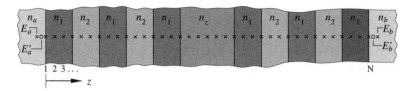

Fig. 15. Longitudinal section of the laser model.

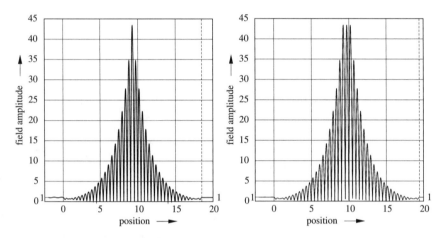

Fig. 16. Electric field distribution in a 1-D laser resonator model.

with $D_x = D_y = 0$. The wave equation is solved by discretization in the z direction. Figure 16 shows the fields in the resonator. We have used the following parameters: $n_1 = 2.0$, $n_2 = 2.5$, $n_z = 2.1$.

The diagram in Figure 17 shows the spectral behavior of the device. The Bragg mirrors consist of 12 Bragg sections and an extra section with $n = n_1$. The intermediate phase-matching section consists of zero, respectively two, periods. The diagram on the right side in Figure 17 shows details of the spectral behavior of the device. It can be seen that the bandwidth decreases with the number of intermediate sections and especially with the number of Bragg sections.

In the VCSEL diode of Figure 14, we have more than 100 layers of different radii. To describe the behavior, we must take into account three different models: the electronic, the optical, and the heat model. These models are coupled with each other, especially by the field-dependent gain in the active layer but also by the heat-dependent parameters of the layers. The solution must be found by an iterative procedure. Furthermore, the procedure is complicated because the eigenresonances must be calculated in the complex plane.

We have calculated in Figure 18 the output power of a special VCSEL diode as function of the injected current. The VCSEL diode was developed at the University of Ulm [24]. As one can see, higher-order modes must be taken

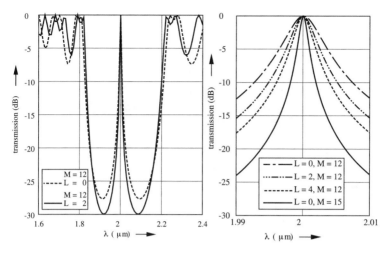

Fig. 17. The spectral behavior of the filter structure in Figure 15, with details in the right diagram.

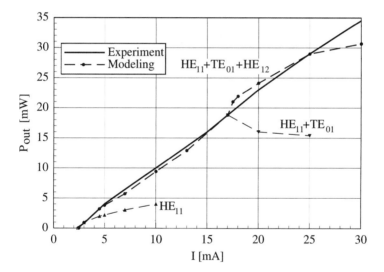

Fig. 18. Emitted power as function of the injected current. (From Ref. [7] © 2001 IEEE.)

into account to obtain results equivalent to the measured results. Each higher-order mode has to be taken into account when the current reaches a special value. The power curve is not continuous. It shows the so-called "kinks." The calculated curve is in good agreement with the measured one. The diagram in Figure 19 shows an intensity plot of the optical field in the VCSEL structure. The radiated field is very small compared to the field in the active region.

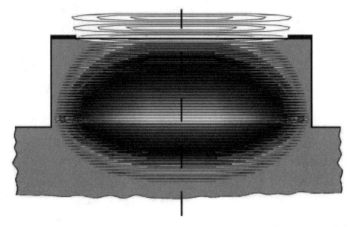

Fig. 19. Intensity plot in the VCSEL diode of Figure 14. (From Ref. [7] © 2001 IEEE.)

Finally, other results not described here should be mentioned. The analysis of a polarization converter is described in Ref. [3]. The analysis of noncircular fibers can be found in Ref. [6].

5 Conclusion

It was shown that efficient algorithms for the analysis of optical devices can be developed on the basis of generalized transmission line equations. These GTL equations standardize and simplify the analysis procedure. With the GTL equations, impedance and admittance transformation formulas can be derived easily. This impedance/admittance transformation is numerically stable. Thus, it avoids numerical problems. Therefore, complex devices can be analyzed. The algorithms were demonstrated and verified with numerical results for eigenmodes, for various devices and especially with results for a VCSEL diode. The analysis follows the physical wave propagation and is therefore efficient.

Acknowledgments

The author would like to acknowledge Stefan F. Helfert and Olaf Conradi for the numerical results in Figure 10 and Figures 18 and 19, respectively.

References

1. Pregla, R., Modeling of planar waveguides with anisotropic layers of variable thickness by the method of lines, Opt. Quantum Electron. 35(4/5), 533–544 (2003).
2. Helfert, S. and Pregla, R., Efficient Analysis of Periodic Structures, IEEE J. Lightwave Technol. 16, 1694–1702 (1998).
3. Helfert, S.F., Barcz, A., and Pregla, R., Three-dimensional vectorial analysis of waveguide structures with the method of lines, Opt. Quantum Electron. 35(4/5), 381–394 (2003).
4. Conradi, O. and Pregla, R., Analysis of fibers with elliptical cross section, 3rd Int. Conf. on Transparent Optical Networks, Cracow, Poland, 18–21 June 2001.
5. Pregla, R. and Conradi, O., Modeling of uniaxial anisotropic fibers with non-circular cross-section by the Method of Lines, 4th Int. Conf. on Transparent Optical Networks, Warsaw, Poland, 21–25 April 2002.
6. Pregla, R. and Conradi, O., Modeling of uniaxial anisotropic fibers with non-circular cross-section by the Method of Lines, IEEE J. Lightwave Technol. 21, 1294–1299 (2003).
7. Conradi, O., Helfert, S.F., and Pregla, R., Comprehensive modeling of vertical-cavity laser-diodes by the Method of Lines, IEEE J. Quantum Electron. 37, 928–935 (2001).
8. Pregla, R. and Pascher, W., The Method of Lines, in *Numerical Techniques for Microwave and Millimeter Wave Passive Structures*, edited by T. Itoh (ed.), Wiley, New York (1989), pp. 381–446.
9. Pregla, R., MoL-BPM Method of Lines Based beam propagation method, in *Methods for Modeling and Simulation of Guided-Wave Optoelectronic Devices*, edited by W.P. Huang, EMW Publishing, Cambridge, MA (1995), pp. 51–102.
10. Pregla, R., Novel FD-BPM for optical waveguide structures with isotropic or anisotropic material, ECIO'99, Torino, Italy, 14–16 April 1999, pp. 55–58.
11. Pregla, R., The impedance/admittance transformation – An efficient concept for the analysis of optical waveguide structures, Integrated Photonics Research Topical Meeting, Santa Barbara, CA, July 1999, pp. 40–42.
12. Felsen, L.B. and Marcuvitz, N., *Radiation and Scattering of Waves*, IEEE Press New York (1996).
13. Pregla, R., Novel algorithms for the analysis of optical fiber structures with anisotropic materials, Int. Conf. on Transparent Optical Networks, Kielce, Poland, June 1999. pp. 49–52.
14. Pregla, R., The Method of Lines as generalized transmission line technique for the analysis of multilayered structures, AEUe Int. J. Electron. Commun. 50, 293–300 (1996).
15. Pregla, R., Efficient modeling of conformal antennas, Millenium Conf. on Antennas and Propagation, Davos, Switzerland, April 2000, paper 0682.
16. Pregla, R., Analysis of planar microwave and millimeterwave circuits with anisotropic layers on generalized transmission line equations and on the Method of Lines, Int. Microwave Symp., Boston, MA, 2000.
17. Pregla, R., Efficient and accurate modeling of planar anisotropic microwave structures by the Method of Lines, *IEEE Trans. Microwave Theory Technol.* 50, 1469–1479 (2002).
18. Pregla, R., Analysis of planar waveguides with arbitrary anisotropic material, Int. Symp. on Electromagnetic Theory, Victoria, Canada, 13–17 May 2001.

19. Pregla, R., Modeling of optical waveguide structures with general anisotrophy in arbitrary orthogonal coordinate systems, IEEE J. Selected Topics Quantum Electron. 37, 1217–1224 (2002).

20. Tripathi, V.K., On the analysis of symmetrical three-line microstrip circuits, *IEEE Trans. Microwave Theory Technol.* 25(9), 726–729 (1977).

21. Čtyroký, J., Helfert, S., and Pregla, R., Analysis of a deep waveguide bragg grating, Opt. Quantum Electron. 30, 343–358 (1998).

22. Greda, L.A. and Pregla, R., Hybrid analysis of three-dimensional structures by the Method of Lines using novel nonequidistant discretization, IEEE MTT-S Int. Microwave Symp., Seattle, WA, 2–7 June 2002.

23. Pregla, R., Efficient modelling of periodic structures, AEUe Int. J. Electron. Commun. 57, 185–189 (2003).

24. Weigl, B., Grabherr, M., Reiner, G., and Ebeling, K.J., High efficiency selectively oxidised MBE grown vertical-cavity surface-emitting lasers, Electron. Lett., 32(6), 557–558 (1996).

Microoptics Using the LIGA Technology

Ines Frese

Institute of Microtechnology Mainz, Carl-Zeiss-Str. 18-20, D-55129 Mainz, Germany
Frese@imm-mainz.de

1 Introduction

The LIGA technology (German acronym for **LI**thography, electroplating (**G**) and molding (**A**) [1,2]) is well suited for the fabrication of complex microoptical systems. This technique allows high-grade integration of microoptical and micromechanical components on the same substrate with a precision in the submicron range. It has the potential for mass fabrication with a variety of optical applications in telecom and datacom approaches, as well as sensing.

A great potential in bandwidth, freedom from electromagnetic interference, and low-cost production are the advantages of glass fibers, which are increasingly replacing electrical data transfer. Glass fiber technology in connection with ultraprecise micromechanical and optical components leads to a completely new possibility for mass production. On the other hand, the LIGA technique allows integrating waveguide and free-space optical components inside one combined component, which can be passively assembled to other elements of the optical setup. A review of the developments of different kinds of microoptical approach investigated at the Institute of Microtechnology Mainz (IMM) demonstrates the advantages of using the LIGA technology for the manufacturing of microoptics.

Whereas this chapter focuses on the LIGA fabrication concept and demonstrates experimental examples for data communications, the subsequent chapter "Modular Fabrication Concept for LIGA-Based Microoptics" by Mohr et al. will be devoted mainly to the presentation of a modular fabrication concept.

2 LIGA Technology for Microoptics

The LIGA technology uses shadow projection lithography via absorber structures. Synchrotron X-rays are used in order to provide sufficient contrast and high steep walls of the developed structures in the resist. These structures are

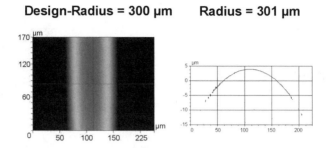

Fig. 1. Contour accuracy of PMMA structures processed by LIGA technology.

filled with metallic negatives using electroplating techniques for producing a mold insert. Finally, this mold insert can be used for mass fabrication, for example, with injection or injection compression molding technology.

2.1 Potential of LIGA technology

For microoptics, LIGA offers certain advantages compared to other technologies. It can provide structures with optical quality as deep as 1 mm, and it opens new ways of supporting monolithic integration of micromechanical and microoptical components. It is possible to fabricate any vertically extruded two-dimensional structure, especially cylindrical microlenses free shaped in one direction (e.g., with aspherical curvature with a lithographical contour accuracy). An example for a contour accuracy is presented in Figure 1.

The lateral resolution of LIGA structures is better than 0.5 µm and aspect ratios of 4 can be realized for complicated geometrical layouts. By using molding techniques, microoptical setups can be fabricated in a variety of optical polymers. Due to the precise pattern transfer from the working mask to the resist via synchrotron radiation, it is possible at the moment at IMM to reach the maximum structural depths limited to 1.3 mm and, at the same time, very steep side walls, less than 100 nm per 100 µm depth. Very important for optical applications is a side wall surface roughness below $\lambda/20$, where λ is the used optical wavelength. Typical surface roughness reached in the LIGA process is below 30 nm [3,4]. This means that the process is useful for optical setups using wavelengths above 600 nm. It is suitable for most telecom and datacom applications.

Wherever there is a demand especially for optical components with high aspect ratios together with optical quality of surfaces, the LIGA process has clear advantages such as the possibility for the following:

- Fabrication of surfaces with optical quality and contour accuracy
- Integration of microoptical and micromechanical features
- Realization of any vertical and inclined cross-sectional shape

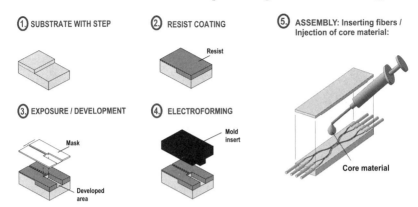

Fig. 2. Scheme of the manufacturing of mold inserts for integrated-optical components with a passive coupling between waveguides and fibers (1-4) as well as for assembly of such a component (5).

Furthermore it is well suited for replication in a variety of optical materials. Such quality of the surface is a result of the use of synchrotron X-rays with high-beam parallelity and energy.

2.2 Microoptical Components and Systems

Advantages of the LIGA technology for the manufacturing of integrated-optical components lie in the possibility to transfer nearly every two-dimensional layout geometry on the resist surface with lithographical precision. After processing with synchrotron radiation and wet-chemical developing, this structure will be extruded perpendicular to the resist surface. It is possible to realize three-level mold inserts by combining precision engineering methods with LIGA technology [5] through insertion of an additional step on the bottom of the structures groove. This step allows a passive vertical alignment between the core of an optical fiber and an optical waveguide because the height of this step is defined through the thickness of cladding of the fiber. This allows one to align axes of the fiber and of the waveguide to each other, promoting a passive coupling between waveguide and fiber suitable for automated manufacturing (Fig. 2).

Dependent on the type of the fibers, a metal or metallized substrate with or without step (Fig. 2, item 1) will be covered with resist (Fig. 2, item 2), which will be depth-lithographically processed (Fig. 2, item 3). After electroforming of the developed resist, a mold insert arises for replication in polymer substrates (Fig. 2, item 4). During assembly, this substrate will be provided with optical fibers, filled with glue with a suitable refraction index as a core material (Fig. 2, item 5), and cured with ultraviolet (UV) light. This manufacturing technique is independent of the type of fibers and is suitable for mass production.

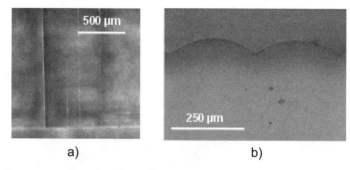

a)   b)

Fig. 3. Some examples of aspherical microlenses produced through the direct irradiation of 1-mm-thick PMMA foil with synchrotron X-rays.

Another example of a LIGA application for microoptics is the realization of cylindrical lenses with spherical or arbitrarily shaped curved surface. The use of a pair of crossed cylindrical microlenses replaces a single spherical microlens because three-dimensional structures like spherical lenses could not be realized directly in the LIGA technology. Two examples of similar combination will be discussed in the following applications.

Due to the projection geometry of the LIGA process, it is impossible to fabricate microlenses in a direct synchrotron irradiation of PMMA foils using a LIGA mask and consecutive etching of irradiated areas. These lenses can be used as separate microoptical components or for the manufacturing of molding tools. In this case, spherical or arbitrarily shaped cylindrical microlenses can be directly fabricated with LIGA with lithographic precision and no further need for posttreatment through the polishing. Figure 3 shows a microscopic photograph (Fig. 3a) and REM photograph (Fig. 3b) of such a one-dimensional aspheric-shaped microlens realized through the direct irradiation at the IMM.

An important application for microlens systems is the imaging of a light source onto a detector using free-space microoptical components. With the help of optical beam propagation simulation programs with sequential (ZEMAX 10.0) and nonsequential ray tracing (OptiCAD 7.0), it can be shown that the distance between source and detector can be dramatically increased when spherical microlenses are replaced by aspherical ones. Figure 4 shows a comparison of simulation results between spherically and aspherically crossed cylindrical microlens systems with a clear-lens aperture of 220 µm. As can be seen, the aspherical approach is superior by a factor of 5 in distance. It makes it possible to realize, for example, an intrachip or interchip optical free-space interconnect between source and detector with a length which can be varied (in the case of this interconnect between 10 and 60 mm). More about this interconnect is discussed in the next chapter.

Another important optical application of LIGA is the production of a molding tool, replication, and assembly of structures with a combined inte-

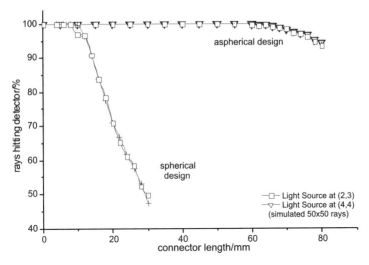

Fig. 4. Simulation and comparison between aspherically and spherically shaped microlens systems: spot size on the detector element at the receiver side versus the propagation length through free space.

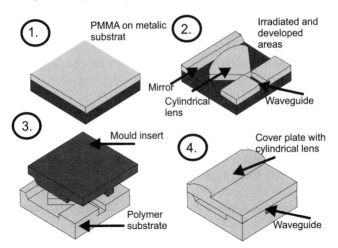

Fig. 5. Fabrication process for a hybrid approach.

gration of waveguide and free-space elements. A fabrication process of such structures including an embedded waveguide in the substrate and free-space 90° outcoupling onto the electric boards is presented in Figure 5. An advantage of such structures is a reduction of the number of separate optical elements requiring an active alignment to the few substrates integrating these components. The substrates are available for a passive alignment to each other. Some concrete structures will be discussed in the next chapters.

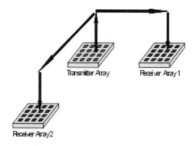

Fig. 6. Interconnection scheme between surface normal transmitter and receiver arrays.

3 Applications

3.1 Free-Space Components: A Refractive Free-Space Microoptical 4×4 Interconnect for Intrachip and Interchip Level with Optical Fan-out

Ongoing demands for higher bandwidth, immunity to electromagnetic interference, and reduction of cross-talk with respect to optical data transport are becoming increasingly important for data rates in the gigahertz domain, especially in optical computation systems. One of the challenges will be the design and fabrication of high-precision microoptical pathway blocks (OPBs) suitable for interchip and intrachip interconnects and for manufacturing at low cost. OPBs should integrate different microoptical components in one optical interconnect between surface-normal transmitter and receiver arrays, as shown in Figure 6 [6–8].

One example of such optical modules was fabricated and monolithically integrated in PMMA (polymethyl methacrylate) by the LIGA technique at IMM [4,9]. The different elements of the optical pathway block are presented in Figure 7. The concept of a refractive free-space microoptical 4×4 interconnect on chip level with optical fan-out was realized. The system is compact, mass producible, and easy to align.

The architecture of an optical pathway block (Fig. 7) is designed for the typical specifications of industry: transmitter and receiver arrays with a pitch of 250 μm in the geometrical arrangement 4×4. The OPB consists of two types of component: microlens array elements (MLEs) to be placed above the transmitter and the receiver arrays and bridging elements (BEs). A bridging element includes microlens arrays crossed to similar ones of MLEs and 45°-tiled micromirrors and establishes an optical path. Both microlens array elements and bridging elements include mechanical holder structures for mounting them onto a separate base plate (BP) with passive alignment. The base plate itself can be mounted on top of the optoelectronic module.

The components of complimentary structures can be transferred into mold inserts, which allow high-volume production in a variety of different polymeric

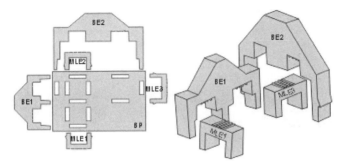

Fig. 7. Two-dimensional and three-dimensional scheme of optical and mechanical elements in the optical pathway concept: bridging elements BE1 and BE2, microlens array elements MLE1, MLE2 and MLE3, and base plate BP.

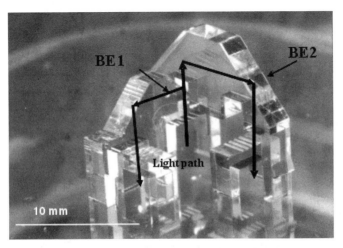

Fig. 8. Photograph of two bridging elements (BE1, BE2) and two microlens array elements (MLE1, MLE2) (setup is without base plate and MLE3). The thickness of the elements is 1 mm. The material is PMMA.

materials. Through the use of LIGA, these free-space optical components were realized with a structural depth of 1 mm with a good optical quality of side surfaces.

A typical roughness measurement of the mirror surface of the bridge value R_a is given by a white-light interferogram with 21 nm over a distance of 1.2 mm. Such an optical quality of the micromirrors integrated into a complicated microoptical component with many different optical elements and available for a low-cost replication as one workpiece without an additional polishing can be reached only in LIGA technology.

A realization of bridging elements and microlens arrays is presented in Figure 8. The elements are designed for an array of 4×4 channels and a maximum interconnection length of 10 mm [9].

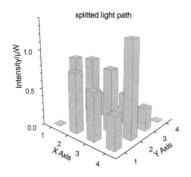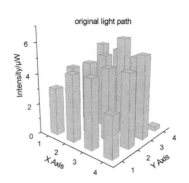

Fig. 9. Measured intensity at detector array 1 (original light path) and detector array 2 (split light path).

In order to realize beam splitting at the interface between BE1 and BE2, the corresponding micromirror at BE1 is covered by sputtering techniques with a 35-nm-thick gold layer that allows an intensity splitting of 50 % in transmission and in reflection, respectively, for a wavelength of 850 nm. The remaining empty space between both interfaces is matched with optical glue that corresponds to the refractive index of the bridging elements.

An individually addressable 4×4 vertical cavity surface emitting laser (VCSEL) array needed for direct proof of principle was not available at that time. That is why light with a wavelength of 633 nm was coupled into the setup of Figure 9 via a single-mode fiber having a numerical aperture (NA) of 0.11 in order to get results on coupling efficiency between transmitter array and both detector arrays. With the help of micropositioners, the fiber was moved to each of the 16 input positions beneath the beam splitter of Figure 4. At both output arrays, the light was collected at the corresponding positions via a multimode fiber with a core diameter of 200 μm and a NA of 0.22.

Figure 9 shows the measured results of the experiment. The intensity of the simulating irradiation from the end of the single-mode fiber was measured to be about 11 μW (without Fresnel reflections). As can be seen, the maximum output energy recorded on the detector was about 6 μW (channel 3,3 in detector array 1).

The bridging and fan-out principles have clearly been demonstrated [9] as well as the direction of the following optimization. Proof of principle for a free-space array interconnect with fan-out is based on the following:

- Integration of microoptical and micromechanical functions
- Crossed cylindrical design
- Refractive fan-out

Free-space interconnection for intrachip and interchip interconnects between optoelectronic arrays will be visible by taking advantage of this LIGA-based approach, allowing a realization of such a highly flexible pick-and-place approach with a high degree of design freedom through the use of aspherical lens shapes. This interconnect is well suited for mass replication in plastics.

3.2 Integrated-Optical Components

In recent years, the need of node components for optical communication networks for datacom approaches suitable for high-speed data transfer up to the data rates of few gigabit per second (Gbps) has become more and more evident [10]. Especially for approaches like automotive or home networks, the need for manufacturing at low cost is an additional demand for a successful transfer of new developments to products. Some examples for node components are optical $N \times M$ couplers and $1 \times N$ splitters, which distribute and route the signals for example in optical backplanes. Developments on the market for optical datacom networks especially for automotive approaches additionally lead to the increasing interest in integrated-optical node components for plastic optical fibers (POFs) or plastic core silica fibers (PCSs). Usually, these elements are demanded in large numbers and therefore require low fabrication costs.

Different types of integrated-optical component were realized at IMM in recent years. Some examples of these are a Y-splitter [11] with excess loss of less than 15 dB and uniformity of better than 0.5 dB and a 4×4 star coupler with excess loss of less than 3 dB and uniformity of better than 2 dB realized for multimode glass fibers [12]. One version of such star couplers is a device for a computing approach with sixfold repeated in the one-plane 4×4 star couplers presented in Figure 10. Some examples for a replicated waveguide structure with a fiber coupling in plastics or electro-formed in an Ni mold insert are presented in Figure 11.

Recent developments for integrated-optical components at IMM concentrate on structures for passive routing for automotive and home network application based on POF and PCS. Some examples for these developments are Y-splitters or 14×14 and 16×16 transmissive star couplers. The main problem of such complicated couplers with a large number of channels is the increase of geometry-based excess losses in comparison with hard requirements of the power budget of optical networks.

Recent investigations in different research centers are carried out now in the direction of excess loss reduction through optimization of geometry and assembly technique. An important tool for that task is the possibilty for quick realization and characterization of components using a rapid prototyping. One of such methods is the Laser-LIGA. In this kind of LIGA technology, a master structure will be realized in a PMMA substrate following production of a mold insert in Ni using electroforming and for the replication in plastics after that. This method was successfully applicated for test structures but is not suitable for direct production of waveguide structures because of a high roughness of the walls, which leads to an increase of excess loss [11]. However, this method

Fig. 10. Photograph of a sixfold 4×4 star coupler as assambled component and packaged device.

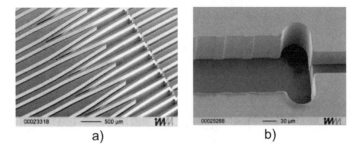

Fig. 11. SEM photographs: (a) the coupling point between a waveguide $(50\times50\,\mu m^2)$ and multimode glass fiber realized in PMMA and (b) a part of an Ni mold insert for a 2×2 coupler array.

allows one to estimate the quality of components with a new geometry without starting a more complicated LIGA processing.

3.3 Components with Combined Free-Space and Integrated-Optical Elements

Increasing data rates by the commercial availability of low-cost active and passive optical components for datacom approaches require new solutions, especially for different kinds of coupling between single optical elements. An overall low-cost approach is needed on both sides for the optoelectronic integration and also for the passive optical interconnects. The main demands are positioning accuracy of the microoptical components, contour accuracy, as well as the possibility for monolithic integration in one substrate of free-space and integrated-optical elements. Especially, the LIGA technology can be helpful in the solution of these demands. One typical task is the effective coupling between electrooptical components and optical fibers suitable

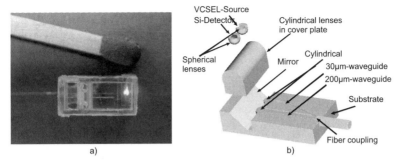

Fig. 12. (a) Photograph of a microoptical unit and (b) principle of coupling scheme of the complete bi-directional transceiver developed at IMM.

for mass production at low cost. The potential of LIGA for such tasks can be demonstrated based on an integrated free-space optical coupling scheme between optoelectronic elements of transceivers and optical waveguides for datacom approaches being developed at the IMM now.

The trend in the development of new optoelectronic one- or bi-directional transceivers lies in the use of surface emitting lasers (VCSEL) and economically favorable silicon detectors. For an effective coupling especially between a silicon detector and an optical fiber, the beam-shaping optical elements are very important. Passive optical interconnects will be based on optical components that can be molded into polymer materials or glass. Monolithic integration in one substrate allows one to reduce the degrees of freedom for critical alignment [5, 13]. An optical unit including two substrates with waveguides, a 45°-tiled micromirror, and two crossed cylindrical microlenses is demonstrated in Figure 12a. The principle of an effective integrated free-space optical coupling scheme between optoelectronic elements of transceivers and optical waveguides is presented in Figure 12b.

The optical setup includes an electrooptical module with a VCSEL source and Si detector provided with spherical microlenses and an optical module. The optical module consists of a base plate, which carried an asymmetrical waveguide circuit, cylindrical microlenses, a tilted 45° mirror, and a cover plate with one asymmetrical common cylindrical lens for both channels. This lens is crossed to both lenses in the base plate. The waveguide circuit includes Y-waveguide splitters with different cross sections of channels. The principle is based on the fact that light from, for example, a surface emitting VCSEL has only a small divergence and can be coupled without significant losses into a waveguide with a smaller cross section ($30 \times 200\ \mu m^2$).

Via a Y-splitter junction, the 30×200-μm^2 waveguide is combined with a waveguide having a cross section of, for example, $200 \times 200\ \mu m^2$. The curved geometry of this waveguide allows one to illuminate the whole cross section of the 200-μm one directly after the junction for the realization of the uniform mode distribution in the waveguide. This is a requirement for a reproductive coupling of the transceiver into the network. Additionally, it makes it possible to separate VCSEL from the back signals coming from the network in the direction of the detector. The straight 200-μm waveguide includes passive coupling of the waveguide to the fiber developed at the IMM [14]. A combination from the first cylindrical lens, a 45°-titled mirror, and the second cylindrical lens crossed to the first one allows collimating of the light from the 200-μm waveguide or focusing of the light into the 30-μm waveguide. This way, a parallel beam of light on the interface between the electrooptical module and cover plate can be received from a source and has tolerances of a few hundred micrometers at the coupling of electrooptical to the optical module at this place. That is why such a coupling scheme can be realized as a solid or a brake-off interconnect.

The main advantage of this setup is given by the use of integration of as many optical elements as possible with the same lithographic technique into one substrate. This makes possible to reduce the number of much separate elements of a complicated optical setup to a few substrates suitable for passive alignment to each other. In our case, the in-plane microlenses, micromirrors, and waveguide grooves have been fabricated with one mask using the LIGA process, with required lateral accuracy in the submicron range. The approach is suited for bus communication in computers or telecom and datacom network applications because the integrated fiber coupling scheme allows one to add other integrated optics components like optical star couplers and can be adapted for arbitrary type of optical fibers.

References

1. Becker, E.W., Ehrfeld, W., Hagmann, P., Maner, A., Münchmeyer, D., Fabrication of microstructures with high aspect ratios and great structural height by synchrotron radiation lithography, galvanoforming, and plastic moulding (LIGA process), Microelectron. Eng. 4, 35 (1986).
2. Ehrfeld, W. and Bauer, H.-D., Application of micro- and nanotechnologies for the fabrication of optical devices, Proc. SPIE 3276, 2 (1998).
3. Brenner, K.-H., et al. Application of three-dimensional micro-optical components formed by lithography, electroforming, and plastic molding, Appl. Optics 32, 6464 (1993).
4. Kufner, M. and Kufner, S., *Micro-optics and Lithography*, VUB Press, Brüssel (1997).
5. Frese, I., Kufner, S., Kufner, M., Hochmuth, G., Bauer, H.-D., and Ehrfeld, W., Combination of guided wave and free-space micro-optics for a new optical backplane concept, Proc. SPIE 4430, 715 (2001).
6. Jahns, J. and Acklin, B., Integrated planar optical imaging system with high interconnection density, Optics Lett. 19(19), 1594 (1993).
7. Verschaffelt, G., et al., Demonstration of a monolithic multi-channel module for multi Gb/s intra-MCM optical interconnect, IEEE Photonics Technol. Lett. 10, 1629 (1998).

8. Becker, E.W., et al., Fabrication of microstructures with high aspect ratios and great structural heights by synchrotron radiation lithography, galvano-forming and plastic molding (LIGA process), Microelectron. Eng. 4, 35 (1986).

9. Voigt, S., Kufner, S., Kufner, M., and Frese, I., A refractive free-space micro-optical 4x4 interconnect on chip-level with optical fan-out fabricated with the LIGA-technique, IEEE Photonics Technol. 14, 1484–1486 (2002).

10. Baierl, W., Robert Bosch GmbH, Evolution of Automotive Networks, POF2001, 2001, p. 161.

11. Klotzbücher, T., Popp, M., Braune, T., Haase, J., Gaudron, A., Smaglinski, I., Paatzsch, T., Bauer, H.-D., and Ehrfeld, W., Custom specific fabrication of integrated optical devices by excimer laser ablation of polymers, Proc. SPIE 3933, 290 (2000).

12. Paatzsch, T., Bauer, H.-D., Popp, M., Smaglinski, I., and Ehrfeld, W., Polymer star couplers for optical backplane interconnects fabricated by LIGA technique, Proc. of Seventh International Plastic Optical Fibre Conference, Berlin, 1998, p. 324.

13. Frese, I., Hochmuth, G., Koch, A., Kufner, S., Nahrstedt, E., Schwab, U., and Voigt, S., Integrated free-space optical coupling scheme between opto-electronic transceivers and optical waveguides, 7th Workshop Optik in der Rechentechnik (ORT), Mannheim, 2002, p. 27.

14. Paatzsch, T., Smaglinski, I., Bauer, H.-D., Ehrfeld, W., Polymer waveguides for telecom, datacom and sensor applications, Miniaturized systems with micro-optics and micromechanics III, Proc. SPIE 3276, 16 (1998).

Modular Fabrication Concept for LIGA-Based Microoptics

Jürgen Mohr, Arndt Last, and Ulrike Wallrabe

Forschungszentrum Karlsruhe, Institute für Mikrostrukturtechnik, Postfach 3640, 76021 Karlsruhe, Germany
Mohr@imt.fzk.de

1 Introduction

Although integrated microoptical systems were in the focus of R&D activities in the past and are still of interest today, it turned out in the last years that especially in the case of sensor applications, a hybrid setup of a microoptical system is a more convenient and economical solution. In this case, it is not necessary to fabricate active devices like lasers and photodiodes, passive components like lenses or waveguides, as well as micromechanical systems like switches or resonators on one substrate by one very specified process technology [1,2]. In a hybrid setup, both the microoptical part, the electrooptical part as well as the mechanical part of the system can be individually optimized.

Whereas technologies for the fabrication of the electrooptical part are already known for years from electronic fabrication and need only to be improved to achieve higher precisions for positioning the various electrooptical devices on a substrate, different microfabrication technologies have been developed and optimized in the past which are able to fabricate microoptical components. Surface micromachining and deep reactive ion etching (RIE) on one side are used to fabricate optomechanical components like switching arrays or movable mirrors in silicon. The LIGA process, on the other hand, is used to fabricate microoptical base plates out of polymers or metals [3].

This concept of microoptical base plates is the reason why the LIGA technology lends itself to the realization of a hybrid integration approach. By using defined interfaces, highly specialized manufacturers can fabricate the different part very efficiently. The equipment costs can be shared among different applications, which makes the component for each application less expensive. Here, a microoptical distance sensor and a microspectrometer for the near infrared (NIR) are used to demonstrate the concept.

This chapter is organized as follows: After a short review of the technology, we will concentrate on the modular design and fabrication of LIGA-based microoptical systems [4].

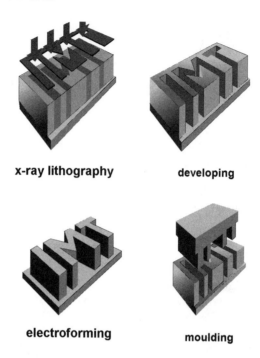

x-ray lithography   **developing**

electroforming   **moulding**

Fig. 1. The process steps of the LIGA process (acronym of the German words Röntgenlithographie, Galvanik, Abformung).

2 The LIGA Process for Optical Applications

To fabricate the microoptical base plate, the LIGA process is used. It was developed at FZK (Forschungszentrum Karlsruhe, Germany) to fabricate polymeric and metallic microstructures [3]. The LIGA process is a combination of lithography methods, electroforming and molding (Fig. 1), which allows one to fabricate microstructures with a high aspect ratio and submicrometer detail. In a first step, a thick polymer layer is patterned by lithography methods. Using X-ray lithography, structural heights up to several millimeters can be achieved with extremely parallel side walls (Fig. 2). Usually, the structures are used as a preform in a subsequent electroforming process to form metallic microstructures or mold inserts. The mold insert is used in either injection molding, hot embossing, or ultraviolet (UV) casting to mass fabricate polymer structures. They have almost the same shape and characteristics as the lithography structures. The extremely parallel side walls, the high precision in structure position, together with the low sidewall roughness makes the process very suitable to fabricate low-cost microoptical components and systems out of polymers, metals, and ceramics (Table 1). Further aspects of the fabrication are discussed in the chapter "Microoptics Using the LIGA Technology" [4].

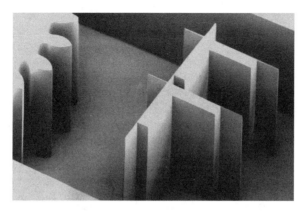

Fig. 2. Typical structure fabricated by the LIGA process (height of the structure: 500 μm; width of the smallest bare: 5 μm).

Table 1. Characteristics of the LIGA Process

aspect ratio > 50	
Precision	Structure details in the submicrometer range
	parallelism of side walls better than 0.4 mrad
Position tolerance	Less than 1 μm over several tens of millimeters
Sidewall roughness	<20 nm root mean square depending on mask quality

The high precision of the LIGA process for optical applications has been demonstrated by two devices: a multifiber connector and a heterodyne receiver.

2.1 Multifiber Connector

The multifiber connector for 16 multimode fibers fabricated by microinjection is illustrated schematically in Figure 3 [5]. It consists of two plastic pieces made of PMMA: one for the alignment of fibers and guide pins with rows of highly precise alignment structures (five are shown in the scheme), the other for their fixation and protection. Elastic ripples in the side walls of the alignment structures facilitate fiber and pin insertion. They also make the connector insensitive to variations in the fiber diameter. The gap between the alignment structures decreases successively from the last row to the first row. This enables a very easy assembly and passive alignment of the fibers without the need of micropositioning. Since the gap at the front face is 2 μm smaller than the fiber diameter, the fibers are clamped softly by the alignment structures, thus allowing easy handling during the ongoing assembly. To bond both parts, UV-curing adhesive is filled into the device through a hole in the upper substrate. Finally, the front face is polished. With these connectors, coupling losses better than 0.5 dB for multimode fibers for all 16 fibers have

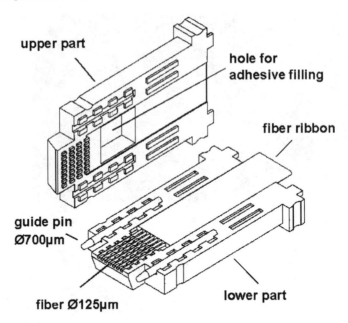

upper part

hole for adhesive filling

fiber ribbon

guide pin Ø700µm

fiber Ø125µm

lower part

Fig. 3. Scheme of a multifiber connector fabricated by microinjection molding using a stepped LIGA mold insert.

been achieved. This demonstrates that the precision of the structures and the structure position is much better than 1 µm.

2.2 Heterodyne Receiver

Figure 4 shows the scheme of the heterodyne receiver (i.e., a wavelength filter for telecommunication). In this case, two incoming light beams need to be split and superposed again [6]. The signal light and the light from a local laser are coupled into the system by means of monomode fibers. The light is collimated by ball lenses and then split in its two polarization states. Reaching the next optical surface, the beam of each polarization state is again split by 50 % and is simultaneously superposed with the respective beam from the opposite light source. Each of the final four superposed beams is detected by a photodiode. The system consists of a ceramic chip on which alignment structures from the polymer are patterned using LIGA technology. Since the optical axis is defined by the diameter of the ball lens (here 900 µm), the fiber with a diameter of only 125 µm needs to be levered on the same height. This is obtained by a precise LIGA fabricated fiber mount. The fibers, the ball lenses, the prisms for the beam splitters, and the diodes are assembled in a fully passive way on the chip. They are just pushed toward the alignment structures and fixed by UV-curing glue. The precision of the alignment structures are better than

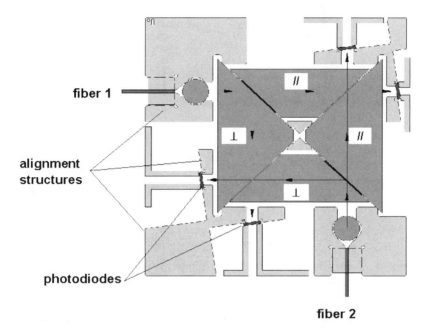

Fig. 4. Scheme of a heterodyne receiver based on a LIGA optical bench.

$1\,\mu m$. This can be concluded from the fact that the superposition of the two beams after the different light paths is better than $95\,\%$.

3 The Modular Fabrication Method

Although the LIGA optical bench allows passive assembly of the individual optical components, the assembly of the heterodyne receiver is very complex and time-consuming. Thus, it is more convenient to follow a modular fabrication concept (Fig. 5). The microoptical system is divided into two functional units: the optical base plate (e.g., an optical bench which is equipped with optical elements like lenses and mirrors) and the electrooptical base plate which covers the active optical elements. If the interfaces between these two components and the qualification methods are well defined, one can fabricate the components separately by using special precise fabrication techniques. By applying a modular concept for the manufacture of the components, cost reduction can be achieved by using the same equipment for different elements.

The two components are combined together to form the microoptical subsystem, which is characterized by its optical output and its electrical interface. This step can be done by a third supplier. The part will be sold as an original equipment manufacturer (OEM) component to the system manufacturer. This supplier that knows the system market very well will combine

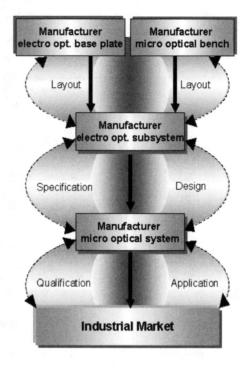

Fig. 5. Modular concept of microoptical system fabrication.

the OEM component with dedicated microelectronics and a customer-defined housing to form the microoptical system. The described fabrication concept is well suited for small- and medium-sized enterprises because each company can make their own profit along the fabrication line. The modular fabrication concept was used to build up an optical distance sensor based on the triangulation principle [7] as well as a microspectrometer [8].

3.1 Distance Sensor

As shown in the scheme of Fig. 6, the microdistance sensor consists of a microoptical and an electrooptical base plate which are assembled together precisely. The dimensions of the sensor chip is 7 mm (W)×7 mm (L)×3 mm (H). The electrooptical base plate is carrying a laser diode, a monitor diode, and a position-sensitive device (PSD) which analyzes the change of the light position detected from the target. The height of the structures on the microoptical base plate is 0.5 mm. So far, the microoptical base plate is fabricated by X-ray lithography using ceramic as a substrate and PMMA as the polymer material (Fig. 7). In the future, it will be fabricated by molding with ceramic as the substrate. The ceramic substrate would have the advantage of a higher temperature stability of the sensor. In Figure 7, the right side shows the illumination optics and the left side shows the detection optics. There are several

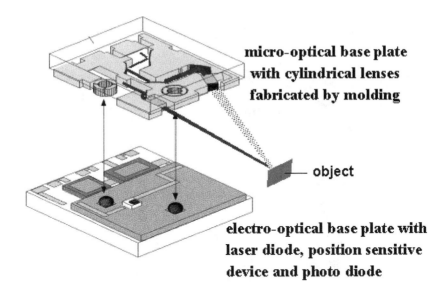

micro-optical base plate with cylindrical lenses fabricated by molding

object

electro-optical base plate with laser diode, position sensitive device and photo diode

Fig. 6. Scheme of the optical distance sensor realized by the modular fabrication concept.

mirrors in the detection part to get an abberration-corrected focus onto the PSD in the horizontal direction. In addition, several light traps are patterned in order to cut out stray light.

To build up the electrooptical subsystem, both base plates are assembled head-over. The alignment is done passively by balls, which are first glued into silicon etch grooves whose position is very precise with respect to the laser diodes position. The balls fit into respective cylindrical tubes fabricated together with the optical structure. By this passive assembly, the laser diode will be positioned into the optical structure with a tolerance of less than 5 µm. Thus, light from the laser diode is coupled precisely into the optical structure and is collimated toward the target. Figure 8 shows the assembled subsystem in a test housing.

Two types of the sensor were designed. The design differs just for the shapes of mirrors and lenses. Distance measurements were carried out with both types of the sensor. The results are summarized in Table 2.

3.2 Microspectrometer for the NIR

A microspectrometer for the NIR is a second example to demonstrate the modular fabrication concept. It consists of an optical base plate which is formed by two parts: the substrate with the spectrometer structure and the waveguide cover (Fig. 9). The substrate of the spectrometer structure together

to the detector

position of the laser diode

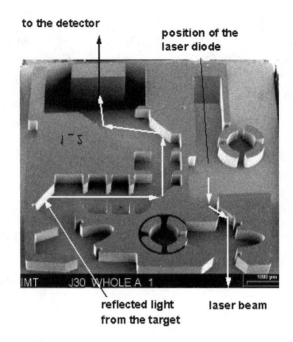

reflected light   **laser beam**
from the target

Fig. 7. SEM photograph of the microoptical base plate with integrated mirrors.

Table 2. Target Specifications and Results of the Micro Distance Sensors

	Measurement range	Working distance*	Linearity	Resolution
Type 1	1 mm	6 mm	$< 0.25\%$	< 1 µm
Type 2	10 mm	16 mm	$< 1\%$	< 10 µm

*The distance between the sensor edge and the center of the measurement range

with the cover provide a "hollow waveguide" to guide the light in the vertical direction by Fresnel reflection. The light is coupled into the waveguide by a fiber and spreads in the horizontal direction freely. The beam hits a cylinder-symmetrical blazed reflection grating which diffracts and reflects the light toward the 45° mirror. It is collimated due to the grating geometry and hits the detector array after being deflected to the vertical direction at the mirror surface.

All of the optical elements are fabricated in one step by hot embossing. The spectrometer substrate and the waveguide cover are assembled passively using some guiding structures in the spectrometer setup. On top of this optical base plate, the electrooptical base plate (FR4 substrate) is mounted. The electrooptical base plate carries the detector array (photo diode) on the front

Fig. 8. Assembled microoptical subsystem of the distance sensor in a test housing.

Table 3. Data of the LIGA Microspectrometers

	Measurement range	Resolution	Stray light compression (full spectrum)
VIS spectrometer	380–760 nm	<7 nm	>17 dB
NIR spectrometer	950–1750 nm	<15 nm	>20 dB

side and, in addition, the analog electronic on the back side. In this case, the alignment between electrooptical and optical base plate is done actively. Figure 10 shows the complete electrooptical subsystem with the analog electronic and the housing. It serves as an OEM component and will be used by spectrometer manufacturers to build up the complete spectrometer system. Microspectrometer systems have been built for the visible range as well as for the NIR range. Table 3 gives the characteristic data for both variations.

4 Conclusion

Because of its unique characteristics, the LIGA process is well suited to fabricate microoptical components and benches. By a modular fabrication process, these benches are combined with electrooptical base plates to build high-quality microoptical systems, as demonstrated by a micro distance sensor and a microspectrometer. Such a modular fabrication concept is very well suited

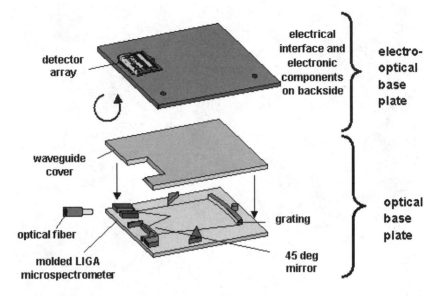

Fig. 9. Schematic drawing to demonstrate the modular setup of the NIR microspectrometer based on a "hollow waveguide."

Fig. 10. Assembled microoptical subsystem of the NIR spectrometer.

in building microoptical sensors in small and medium numbers. It allows also small- and medium-sized sensor manufacturers to make use of the advantages of microfabrication technology. In the future, the concept will be applied to demonstrate advanced sensor systems. In collaboration with companies, an industrial fabrication line will be established.

References

1. Marxer, C. and de Rooij, N.F., Micro-opto-mechanical 2x2 switch for single mode fibers based on a plasma-etched silicon mirror and electrostatic actuation, J. Lightwave Technol. 17(1), 2–6 (1999).
2. Wu, M.C., Lin, L.Y., Lee, S.S., and King, C.R., Free-space integrated optics realized by surface-micromachining, Int. J. High Speed Electron. Syst. 8(2), 283–297 (1997).
3. Mohr, J., LIGA – A technology for fabricating microstructures and microsystems, Sensors Mater. 10, 363–373 (1998).
4. Frese, I.: Microoptics using the LIGA technology, this book
5. Wallrabe, U., Dittrich, H., Friedsam, G., Hanemann, Th., Müller, K., Piotter, V., Ruther, P., Schaller, Th., and Zissler, W., Micromolded easy-assembly multi fiber connector RibCon, Microsyst. Technol. 8, 83–87 (2002).
6. Ziegler, P., Wengelink, J., and Mohr, J., Passive alignment and hybrid integration of active and passive optical components on a microoptical LIGA bench, Proc. 3rd Int. Conf. On Micro Opto Electro Mechanical Systems, MOEMS'99, Mainz, 1999, pp. 186–189.
7. Oka, T., Nakajima, H., Tsugai, M., Hollenbach, U., Wallrabe, U., and Mohr, J., Development of a micro-optical distance sensor, Sensors Actuators, 102, 261–267 (2003).
8. Krippner, P., Kühner, T., Mohr, J., and Saile, V., Microspectrometer system for the near infrared wavelength range based on LIGA technology, Proc. SPIE 3912, 141–149 (2000).

Fabrication of Optical Waveguides in Dielectrics by Femtosecond Laser Pulses

Stefan Nolte, Matthias Will, Jonas Burghoff, and Andreas Tünnermann

Institute of Applied Physics, Friedrich-Schiller-University Jena, Max-Wien-Platz 1, 07743 Jena, Germany
nolte@iap.uni-jena.de

1 Introduction

Modern communication systems are based on integrated optical devices to control the properties of light in all-optical networks. Key elements within these networks are active and passive waveguides, splitters, connectors, and filters. The optical function of these elements is based on a spatial refractive-index modification within the glass matrix, which is typically fabricated by ionic exchange or diffusion processes. Although these technologies are well established and very successful, their application is, in general, restricted to the generation of planar (two-dimensional) elements.

In recent years, a novel technique based on the use of ultrashort laser pulses for the direct writing of photonic structures within different glasses [1–22] and also crystalline media [23] has been demonstrated. When tightly focused into the bulk of a transparent solid, femtosecond laser pulses can produce a permanent refractive-index modification. The laser energy is absorbed in the focal volume by multiphoton and avalanche absorption, leading to optical breakdown and the formation of a microplasma [24–26]. This induces permanent structural and refractive-index modifications three-dimensionally localized inside the bulk substrate. When the samples are moved with respect to the laser beam focus a refractive-index profile can be generated that allows one to guide light. This enables the direct fabrication of buried optical waveguides and more complex three-dimensional photonic devices with great flexibility in different transparent media.

The physical mechanisms that are responsible for the refractive-index changes are not completely understood, so far. Although, for example, in fused silica microstructural changes are leading to densification [11, 21], in crystalline media, induced stress seems to play an important role [23]. Despite this, several optical devices have already been demonstrated (e.g., optical waveguides [1, 4, 8, 13, 15, 18, 20, 21], waveguide amplifiers [9, 17], beam splitters [5], X-couplers [12, 16], directional couplers [14], stacked waveguides and waveguide arrays [12], three-dimensional waveguiding elements [22], three-dimensional data storage [2], transmission gratings [7], and long-period fiber gratings [6]).

Here, we report on our recent advances in femtosecond direct writing of photonic devices. The properties of waveguides fabricated in glasses and crystals and their dependence on the production laser parameters as well as processing requirements will be discussed.

2 Experimental

Refractive-index changes are generated using a commercial amplified Ti:sapphire femtosecond laser system (Spectra-Physics, Spitfire). It produces 50-fs pulses with a repetition rate of 1 kHz at a wavelength of 800 nm. To produce the localized refractive-index changes, pulses with an energy of ~ 1 μJ were focused into the transparent samples approximately 200 μm below the surface. The focusing is accomplished either by a 10× microscope objective with a numerical aperture (NA) of 0.25 or by a 20× microscope objective (NA = 0.45), which is corrected for a focal depth of 170 μm.

In a first approximation, the transversal and longitudinal structure sizes produced can be estimated from the focus diameter and confocal parameter, respectively. However, the actual dimensions might differ slightly from these predictions. Due to the nonlinear absorption process, a reduction of the modified region can be expected. On the other hand, spherical aberration will lead to slightly larger structures.

Optical waveguides are fabricated by moving the samples perpendicular to the laser-beam axis, as shown in Figure 1, by a computer-controlled three-axis positioning system (Physik Instrumente, M-126.DG) with a maximum velocity of 1.5 mm/s. The third axis is used for controlling the focal depth inside the target. Using this setup, waveguides as long as 5 cm have been written, which was limited only by the travel range of the positioning system.

Although this setup allows one to write waveguides with great flexibility, it has the disadvantage that the cross section of the waveguides depends strongly on the focusing conditions. As a consequence, the waveguides are elliptical due to the mismatch between focus radius and Rayleigh length, if no objectives with high NAs (close to 1) are used. However, this would cause problems due to spherical aberration, especially when focusing in different depths. This can be overcome by shaping the beam [15] or by writing the waveguides along the laser-beam axis. In the latter case, however, the length of the waveguides as well as the flexibility are limited.

In order to characterize the optical function of the produced devices, the sample surfaces (entrance and exit side) are polished after laser processing. Laser radiation at different wavelengths is coupled into the waveguides via single-mode fibers that are butt-coupled to the sample surface. The near-field intensity distribution of the guided modes is obtained by imaging the

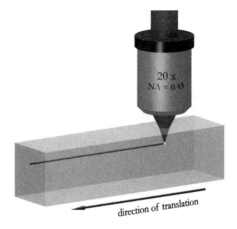

Fig. 1. Schematic of the experimental setup.

exit surface onto a charge-coupled device (CCD) camera by a microscope objective. In case the produced waveguide is multimodal, different propagation modes can be selectively excited by careful adjustment of the fiber–waveguide coupling.

Direct measurements of the refractive-index profiles of the written waveguides are performed using a commercial refracted-near-field (RNF) profilometer (Rinck Elektronik, Germany). The operation principle can be found in [27]. Absolute values of the refractive index are obtained by repeating measurements with a reference sample and an immersion fluid with well-known refractive indices. Although the refractive-index measurements are performed at a wavelength of 635 nm, we assume that the produced relative-index changes are approximately the same for other wavelengths as well.

Based on the measured refractive-index profiles, calculations of the field distribution of the guided modes are performed. For this purpose, the commercial program BeamPROP 5.0 (Rsoft, Inc.) is used, which solves the Helmholtz equation in the paraxial approximation using a finite-difference beam-propagation method.

In addition to the optical characterization of the produced structures, modifications generated in crystalline materials are analyzed with respect to their morphological structure using X-ray topography and transmission electron microscopy (TEM).

3 Waveguides in Glasses

Figure 2 shows the near-field intensity profile of a straight waveguide, which has been written 200 μm below the surface in a fused silica substrate with a pulse energy of $\sim 1\,\mu J$ at a writing speed of 100 μm/s. The focusing was

128   Stefan Nolte et al.

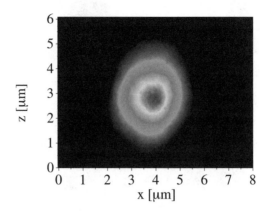

Fig. 2. Near-field intensity distribution at the end of a straight waveguide (for a wavelength of 800 nm).

realized by a 20×, NA = 0.45 microscope objective. For obtaining the near-field distribution, light at a wavelength of 800 nm was launched into the waveguide by butt-coupling a fiber to the polished-end surface, and the exit surface of the waveguide has been imaged onto a CCD camera. Figure 2 indicates that the produced waveguide is single mode at the wavelength of 800 nm.

The induced refractive-index modifications are very stable; the waveguiding properties are preserved after heating the samples up to 500°C for several hours. In contrast, the characteristic fluorescence of color centers, which are known to be formed during the waveguide fabrication process [1, 19, 21], disappears when the sample is heated to ~ 400°C. After this annealing process, no fluorescence emission is observable, and the near-field distribution of the guided light has not changed [20]. These observations, which are in agreement with those reported in Ref. [21], allow one to conclude that the refractive-index modifications in fused silica and the waveguiding properties are not determined by the generation of color centers.

Important for applications of this technology are the losses light experiences when propagating through the structures. In order to determine the attenuation losses, we applied the so-called cut-back technique, where a waveguide is cut into samples of different length. The total transmission through each of these samples is measured after the surfaces have been polished. By comparing the transmission through the different samples, the coupling losses as well as the damping losses can be determined independently. At a wavelength of 633 nm, we obtained damping losses of below 0.4 dB/cm.

In addition to, single-mode waveguides, multimode waveguides with a defined number of guided modes can be produced simply by changing the writing speed (i.e., the spatial overlap of successive pulses). A lower writing speed and, thus, a higher number of pulses results in a stronger increase of the refractive-

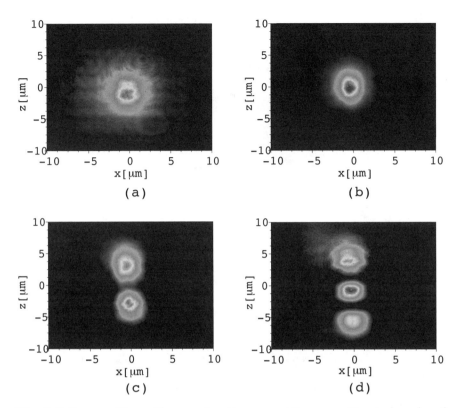

Fig. 3. Influence of the writing speed on the waveguiding properties at a wavelength of 514 nm. Only the highest-order guided modes are shown for a writing speed of (a) 1000 μm/s, (b) 500 μm/s, (c) 80 μm/s, and (d) 25 μm/s.

index change. As a consequence, the near-field distribution is different. In Figure 3, the measured near-field intensity distributions of waveguides written at different writing speeds are shown. In all cases, only the highest-order mode for 514-nm laser radiation that is guided in these structures is shown. In this case, the focusing is done with a 10×, NA = 0.25 microscope objective, and a pulse energy of 1 μJ is used. At writing speeds of 1000 μm/s (Fig. 3a) and 500 μm/s (Fig. 3b), the femtosecond-laser-induced refractive-index changes are small, and only single-mode operation can be observed. If the writing speed is decreased down to 80 μm/s (Fig. 3c) and 25 μm/s (Fig. 3d), the refractive-index increase is large enough that higher-order modes are guided (in the horizontal axis, the waveguides are still single mode due to the smaller diameter in this direction compared to the vertical axis). RNF measurements yield a refractive-index increase of about $\Delta n \approx 5 \times 10^{-4}$ for a writing speed of 1000 μm/s. In case of the slow writing speed (25 μm/s), the measured refractive-index increase amounts to $\Delta n = 3 \times 10^{-3}$.

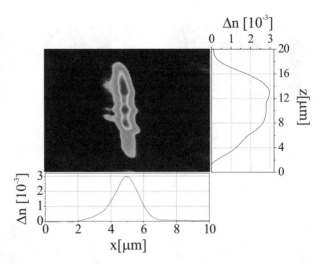

Fig. 4. Measured refractive-index profile of the waveguide written at 25 μm/s.

Figure 4 shows the refractive-index profile (cross section, the writing laser beam was incident from the top) of the waveguide written at a speed of 25 μm/s measured at a wavelength of 633 nm using the RNF profilometer. The horizontal and vertical profiles show approximately a Gaussian shape and fit very well to the dimensions of the focus diameter and confocal parameter, respectively. Based on this refractive-index profile, the field distribution of the guided modes was calculated using the commercial program BeamPROP 5.0 (RSoft, Inc.). The calculated intensity distribution for the highest-order guided mode at a wavelength of 514 nm is shown in Figure 5a. For comparison, the measured near-field intensity profile of the highest-order mode guided in the corresponding waveguide is shown in Figure 5b. The measured and calculated intensity distributions are in very good agreement.

It is important to note that even the waveguide written at a speed of 25 μm/s shows single-mode operation when a larger radiation wavelength around 1.5 μm is used, which is important for optical communications. In this case, the calculated and measured near-field intensity distributions are in good agreement as well [20].

For the fabrication of more complex devices than straight waveguides, the realization of bends is important. While sharp bending angles are known to produce high losses, the bending can be realized by so-called S-bends (see Fig. 6a). In this case, the light is guided along a smooth curve with minimal radius and without discontinuities in the direction of the waveguide. In order to evaluate the amount of lateral displacement that can be achieved, we produced several S-bends with different lateral displacements and a total length of 10 mm at a writing speed of 100 μm/s using different pulse energies. For

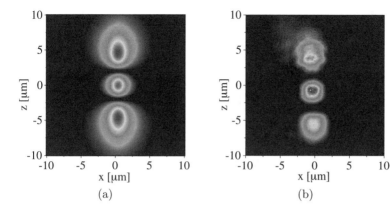

Fig. 5. Comparison of (a) calculated and (b) measured guided light intensity distributions at 514 nm for the waveguide written at 25 μm/s (only the highest-order guided mode is shown). For the calculations, the measured refractive-index profile of Figure 4 was used.

the focusing, a 20×, NA = 0.45 microscope objective was used. In Figure 6b, the resulting values of the relative transmission through these devices at a wavelength of 514 nm are marked as data points for pulse energies of 0.13 μJ and 0.5 μJ. As can be seen, by using pulse energies in excess of 0.5 μJ, lateral displacements of more than 100 μm can be achieved for a device length of 10 mm without significant losses.

To compare the experimental data with simulations, the refractive-index distribution of a waveguide (written with a pulse energy of 1 μJ in this case) was measured (Fig. 7). The simulations were performed using the shape of this profile with different values of the maximum refractive-index change for S-bends written perpendicularly to the laser beam axis. The results are summarized in Figure 6b for different lateral displacements. A good agreement with the experimental data for pulse energies of 0.13 μJ and 0.5 μJ is obtained when maximum refractive-index values of 3×10^{-4} and 8×10^{-4}, respectively, are used for the simulations.

Despite of all the planar structures discussed so far, one of the main advantages of this technology is the possibility to open the door to 3-D integrated optics, which allows one to increase the packaging density of optical functions tremendously. In order to obtain the necessary information to fabricate true 3-D elements, the same simulation for the S-bend displacement has been performed along the other axis (parallel to the laser beam axis), since the femtosecond laser written waveguides are not circularly but elliptically shaped (see Fig. 7). However, due to the larger extension of the modified refractive-index in this direction, the damping losses are much less sensitive to the amount of S-bend displacement.

S-bend distance

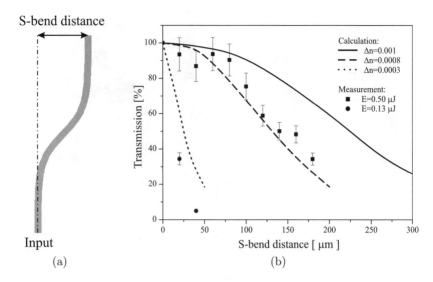

Input

(a)   (b)

Fig. 6. (a) Schematic of an S-bend. The lateral offset (S-bend distance) is varied, and the transmission is calculated. The total device length is 10 mm. (b) Dependence of the transmitted light intensity (514 nm) on the lateral S-bend distance. The lines correspond to calculations based on different refractive-index changes Δn, whereas the squares and circles show experimental data points normalized to a straight waveguide (S-bend distance = 0).

With these data, all necessary information for the realization of a true three-dimensional integrated optical device is available. To demonstrate this possibility, a 1×3 splitter was fabricated. For this purpose, the positioning system was programmed to write the split arms under angles of 120°. This means that none of the exit ports is lying in the same plane as the entrance arm. In order to keep the losses minimal, the sample was moved with respect to the laser-beam axis and the focusing objective in such a way that the single arms are formed as S-bends with only small bending radii. The refractive-index changes are induced by $\sim$ 1-µJ pulses focused with a 20× microscope objective (NA = 0.45) approximately 200 µm below the surface of a fused silica sample. The writing speed was 125 µm/s, and the total length of the device was programmed to be 10 mm with the exits separated by 100 µm/s. Figure 8 shows a schematic of the production process and the resulting device.

To measure the guided light intensity distribution, laser radiation at 1.05 µm of a fiber-coupled laser diode was coupled into the waveguide splitter. The exit surface of the sample was imaged onto a CCD camera. Figure 9a shows the near-field distribution of the guided light. The splitting ratio (32 : 33 : 35) is almost equal. In this case, the entrance port of the splitter was slightly multimode, whereas the exits showed single-mode behavior. However, as discussed earlier, the guiding properties (single-mode or multi-

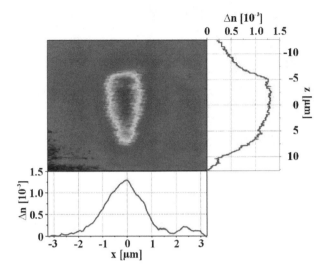

Fig. 7. Measured refractive-index distribution of a femtosecond written waveguide in fused silica (20× focusing objective (NA = 0.45), pulse energy 1 μJ).

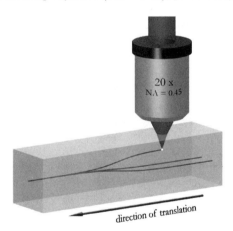

Fig. 8. Schematic of the writing process of a three-dimensional 1×3 splitter.

mode behavior) can simply be determined by adjusting the writing speed. The splitting losses have been determined to ∼ 6 dB by measuring the overall transmission losses and comparing these data to the propagation losses through a straight waveguide written with the same parameters. A part of the splitting and damping losses can surely be attributed to the imperfections (positioning error, vibrations, etc.) of the positioning system used. However, the actual amount is still under evaluation.

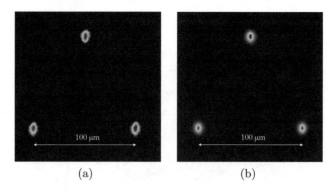

(a)   (b)

Fig. 9. (a) Near-field intensity distribution at 1.05 μm measured at the exit of the splitter. The splitting ratio is 32 : 33 : 35. (b) Calculated near-field intensity distribution at the end of the splitter. The calculations are based on the measured refractive-index profile (Fig. 7).

Based on the measured refractive-index distribution of Figure 7 (approximated by Gaussian distributions), the guiding and splitting properties of the three-dimensional device were calculated. Figure 9b shows the calculated intensity distribution at a wavelength of 1.05 μm at the end of the waveguide splitter, which agrees well with the measured near-field distribution of Figure 9a.

Important for the production of integrated optical components based on the femtosecond direct-writing technique are the requirements on the accuracy of the positioning system. Thus, we simulated the required positioning accuracy based on the measured refractive-index distribution (Fig. 7) for a Y-splitter based on S-bends. The layout of the Y-splitter with a total length of 10 mm is shown in Figure 10a. As is obvious from this schematic, a positioning error is taken into account by a lateral offset of the second arm from the ideal branching point. For different positioning errors (lateral offsets), the output power distribution in the two exit ports has been calculated for a wavelength of 514 nm. The results are summarized in Figure 10b for a Y-splitter produced in the plane perpendicular to the laser beam axis (i.e., within the plane of the short axis of the elliptical refractive-index distribution) (the production requirements for a Y-splitter along the other axis are more relaxed, since the region of overlap is much broader in this case). For a negative lateral offset, the two exit ports are crossing and the splitting ratio is changing rapidly between the two arms. For positive positioning errors, the two exit ports are separated from each other without any crossing point. Thus, the power in the right arm is decreasing with increasing offset, since light can no longer couple into this arm. From these data, it is obvious that the positioning system must provide a positioning accuracy of at least ±1 μm in order to produce a Y-splitter with a defined splitting ratio, which is feasible with commercial positioning systems.

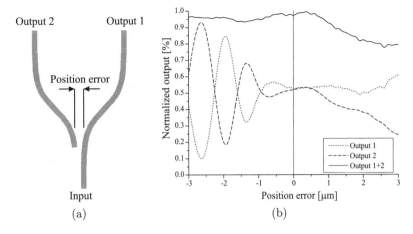

Output 2   Output 1

Position error

Input

(a)

Position error [μm]

(b)

Fig. 10. Influence of the positioning error on the output power distribution of a Y-splitter. (a) Schematic of the Y-splitter with a lateral offset of the second arm from the ideal branching point. (b) Simulation of the output power distribution at a wavelength of 514 nm as a function of the positioning error (lateral offset). To achieve an approximately equal splitting ratio, the positioning accuracy has to be better than ±1 μm.

Despite of the possibility to produce real 3-D integrated optical elements, the ultrashort-pulse direct-writing technique has one additional advantage that is important for the fabrication of complex high-density integrated-optical devices: It can be applied to a large variety of different materials. As an example, Figure 11 shows a phosphate glass sample (Schott IOG-1) with an undoped area and an area doped with Er/Yb. The sample was produced by bonding the two phosphate glasses together [28]. Using the femtosecond laser, we have written a waveguide through the bond. Identical writing parameters were used for both sides. In Figure 11, only the fluorescence of light at 543 nm in the doped part of the glass is visible, whereas the pump light at 800 nm is coupled into the sample from the other side (through the undoped region). In combination with the 3-D structuring possibilities, this demonstrates the potential to realize complex devices using the femtosecond direct-writing technique.

4 Waveguides in Crystalline Media

In addition to the fabrication of photonic devices in glasses, the femtosecond direct-writing technique has the potential to produce waveguides in crystalline materials. Figure 12 shows a polarization contrast optical microscope image of a cross-cut through a waveguide in crystalline α-quartz, which has been produced with the same setup as used for the fused silica samples but at a

Fig. 11. Femtosecond written waveguide in a bonded glass with a doped (Er/Yb) and an undoped region. Pump light at 800 nm is coupled into the undoped region. The fluorescence from the doped region demonstrates that the waveguide was successfully written through the bond.

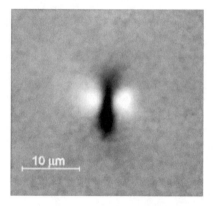

Fig. 12. Polarization contrast optical microscope image of a waveguide (cross section) written in crystalline α-quartz with femtosecond laser pulses.

slightly higher pulse energy (14 μJ, writing velocity 0.1 mm/s, focusing 10×, NA = 0.25 microscope objective).

In Figure 12, two different features are noticeable. The dark central area corresponds to the focal region, where the laser radiation was absorbed (the laser beam was incident from the top of the image), resulting in increased scattering losses. This dark area is surrounded by two bright regions, indicating a refractive-index change. The maximum refractive-index increase Δn in these regions was estimated from interferometric analysis to be $\Delta n \approx 0.01$. Due to the increased refractive-index in these bright areas, light can be guided in the produced structure, as the near-field intensity distribution of 514-nm radiation in Figure 13 shows. Corresponding to the refractive-index distribution,

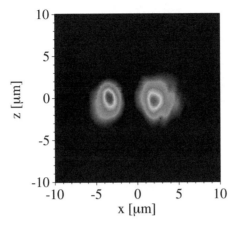

Fig. 13. Near-field distribution of 514-nm radiation guided in the waveguide produced by femtosecond laser pulses in crystalline α-quartz.

the light is guided not in the central part, but in the areas showing up bright in Figure 12. Note that the near-field distribution does not show a higher-order mode as in Figure 5b (in that case, the peaks would have to be oriented along the other axis since the Rayleigh length is larger than the beam radius).

In order to reveal the induced microstructural modifications that lead to the refractive-index changes, TEM studies of the modified area were performed. For this purpose, thin slices have been cut from the laser-irradiated crystals parallel to the beam direction, perpendicular to the written waveguide. These slices were thinned using mechanical polishing, dimpling, and ion milling.

Figure 14 shows a cross-sectional TEM image of the modified region. The irradiated area has an elliptical shape with a width of about 1 μm and a height of approximately 10 μm. Three different regions can be identified as indicated in the image. Whereas region I has a high density of defects but the material is still crystalline, the central part II is amorphous. This has been proven by taking selected-area electron diffraction (SAED) pattern images. The third region (III) is crystalline again but has a strong strain contrast.

The focal area, where the laser light has been absorbed, is surrounded by a region showing high internal strains in the matrix. This can be deduced from the numerous bending contours present in this area. Since it is very reasonable that this strong strain field is responsible for the refractive-index increase, we simulated the strain field after single-shot exposure by the finite-element method (FEM) using the code ANSYS. Possible recrystallization effects have been neglected. As input, we used an expanding ellipse with a diameter ratio of 1 : 10 embedded in a crystalline quartz matrix.

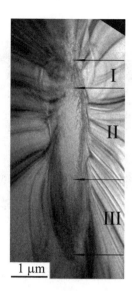

1 μm

Fig. 14. Cross-sectional TEM image of a waveguide written with femtosecond laser pulses. Region I has a defective crystalline structure, region II is amorphous, and region III shows a complex strain contrast of the crystalline matrix.

A two-dimensional solution was obtained within isotropic elastic approximations using the elastic constants of fused silica ($E = 72\,\mathrm{Gpa}$, $\nu = 0.17$) [29]. The inner phase was simulated with a fictive thermal expansion coefficient α of 0.06 and a fictive temperature change ΔT of unity. The value $\alpha \cdot \Delta T$ reflects the length change from crystalline to amorphous quartz according to the density change of the bulk quartz matrix during amorphization [30]. Figure 15 displays the hydrostatic pressure distribution given by the trace of the strain tensor according to the FEM results. Since the simulation does not reflect the actual behavior of the inner part, the solution is not valid inside the irradiated ellipsoidal region. However, two maxima of the strain distribution are clearly visible on both sides of the central ellipse. These maxima reflect a local increase in the material density, to which the refractive-index is proportional in a first approximation. These results are in qualitative agreement with the polarization contrast microscope image (Fig. 12), which shows a similar distribution.

The results of the TEM analysis are confirmed by X-ray topography measurements. These measurements yield a strongly disturbed core with an area of approximately $1 \times 10\,\mu\mathrm{m}^2$ (in agreement with the TEM analysis). After a short transition zone of $\sim 50\,\mathrm{nm}$, this central core is surrounded by a deformed crystalline lattice.

The measured losses of the waveguides in crystalline quartz are lower than $5\,\mathrm{dB/cm}$. This is significantly higher than in glass, which is probably due to the disturbed central area. However, the produced refractive-index changes

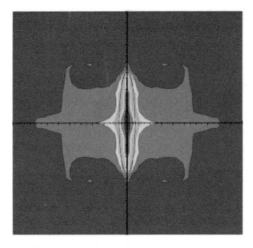

Fig. 15. Distribution of the hydrostatic pressure in the quartz matrix after irradiation with one femtosecond laser pulse (simulated as an expansion of the central, elliptical area) according to the FEM calculations. Due to the simplifications the solution is not valid in the central area.

are very stable. They remain present even after thermal annealing at 1200°C for several hours.

5 Conclusion and Outlook

In conclusion, optical waveguides have been fabricated by femtosecond laser pulses inside glasses and crystalline quartz. In fused silica, we demonstrated refractive-index changes up to $\Delta n = 3 \times 10^{-3}$ and waveguides with losses below 0.4 dB/cm. It has been shown that by changing only the writing speed, it is possible to produce waveguides with a defined number of guided modes. For a laser wavelength of 514 nm, both single-mode and multimode waveguides have been fabricated. The measured near-field intensity distributions were in good agreement with simulations based on the measured refractive-index profiles.

In addition to these straight waveguides, true three-dimensional photonic devices have been produced. As a demonstration, a 3-D 1×3 splitter with a total length of 10 mm and arms split by 100 μm has been fabricated in fused silica using ultrashort laser pulses.

Another advantage of this technique is that it is not limited to a certain material like fused silica. Different types of glass, doped and undoped, can be structured as well as crystalline media. In crystalline quartz, waveguides with damping losses below 5 dB/cm have been realized. The achieved refractive-index changes are up to $\Delta n \approx 0.01$ and are stable up to temperatures above 1200°C. In this case, the refractive-index modifications are due to strong strains in the crystalline matrix arising from an amorphization of the irradiated area, as micromorphological TEM and X-ray analyses have revealed. The light is guided not in the amorphous but in the crystalline region of the material. The possibility to produce waveguides in crystalline media makes this technique a promising candidate for integrated optical frequency conversion and electro-optical switching applications.

The femtosecond direct-writing technique thus allows for rapid prototyping of complex three-dimensional photonic elements in various materials. The main drawbacks so far, the limited refractive-index changes and the writing speed in the range of 1 mm/s, could be overcome by using higher-repetition-rate lasers. Whereas modified conventional femtosecond oscillators are producing barely enough pulse energy to induce the required changes, new developments in laser sources allow pulse energies in the order of 1 μJ at repetition rates in the megahertz range [31,32]. This enables flexible structuring at high speed, which makes this technology even suited for mass production. In a first experiment with a fiber-based laser system producing 1-μJ, 300-fs pulses at a repetition rate of 2 MHz, we were able to write waveguides in fused silica at a speed of 100 mm/s, limited only by the positioning system. The waveguides show a refractive-index increase of the order of 10^{-2}. Preliminary measurements of the attenuation losses yielded values of the order of 0.5 dB/cm.

It is important to note that at such high repetition rates, the time between successive pulses is not long enough to allow the deposited heat to diffuse away before the next pulse arrives. As a consequence, localized melting inside the solid occurs due to the accumulative heating of several following pulses [13,18,19]. The material resolidifies when the laser beam is moved to another place. It will be very interesting to see how this technology develops in the near future.

Acknowledgments

We gratefully acknowledge the assistance of T. Gorelik and U. Glatzel for the TEM investigations and the simulation of the strain field, F. Wunderlich and K. Goetz for preparing the X-ray analysis, and D. Blömer for the measurement of the waveguide losses. Furthermore, we would like to thank M. Brinkmann from SCHOTT Glas, Germany, and J. S. Hayden from SCHOTT Glass Technologies, Inc., USA, for useful discussions and the provision of the bonded glass samples. This work was supported by the Deutsche Forschungsgemeinschaft (SFB 196) and the Thuringian Ministry of Science, Research and Art (TMWFK, B507-02006).

References

1. Davis, K.M., Miura, K., Sugimoto, N., and Hirao, K., Writing waveguides in glass with a femtosecond laser, Opt. Lett. 21, 1729–1731 (1996).

2. Glezer, E.N., Milosavljevic, M., Huang, L., Finlay, R.J., Her, T.-H., Callan, J.P., and Mazur, E., Three-dimensional optical storage inside transparent materials, Opt. Lett. 21, 2023–2025 (1996).

3. Glezer, E.N. and Mazur, E., Ultrafast-laser driven micro-explosions in transparent materials, Appl. Phys. Lett. 71, 882–884 (1997).

4. Miura, K., Qiu, J., Inouye, H., Mitsuyu, T., and Hirao, K., Photowritten optical waveguides in various glasses with ultrashort pulse laser, Appl. Phys. Lett. 71, 3329–3331 (1997).

5. Homoelle, D., Wielandy, S., Gaeta, A.L., Borrelli, N F., and Smith, C., Infrared photosensitivity in silica glasses exposed to femtosecond laser pulses, Opt. Lett. 24, 1311–1313 (1999).

6. Kondo, Y., Nouchi, K., Mitsuyu, T., Watanabe, M., Kazansky, P G., and Hirao, K., Fabrication of long-period fiber gratings by focused irradiation of infrared femtosecond laser pulses, Opt. Lett. 24, 646–648 (1999).

7. Sudrie, L., Franco, M., Prade, B., and Mysyrowicz, A., Writing of permanent birefringent microlayers in bulk fused silica with femtosecond laser pulses, Opt. Commun. 171, 279–284 (1999).

8. Korte, F., Adams, S., Egbert, A., Fallnich, C., Ostendorf, A., Nolte, S., Will, M., Ruske, J.-P., Chichkov, B.N., and Tünnermann, A., Sub-diffraction limited structuring of solid targets with femtosecond laser pulses, Opt. Express. 7, 41–49 (2000).

9. Sikorski, Y., Said, A.A., Bado, P., Maynard, R., Florea, C., and Winick, K.A., Optical waveguide amplifier in Nd-doped glass written with near-IR femtosecond laser pulses, Electron. Lett. 36, 226–227 (2000).

10. Watanabe, W., Toma, T., Yamada, K., Nishii, J., Hayashi, K., and Itoh, K., Optical seizing and merging of voids in silica glass with infrared femtosecond laser pulses, Opt. Lett. 25, 1669–1671 (2000).

11. Chan, J.W., Huser, T., Risbud, S., and Krol, D.M., Structural changes in fused silica after exposure to femtosecond laser pulses, Opt. Lett. 26, 1726–1728 (2001).

12. Minoshima, K., Kowalevicz, A.M., Hartl, I., Ippen, E.P., and Fujimoto, J.G., Photonic device fabrication in glass by use of nonlinear materials processing with a femtosecond laser oscillator, Opt. Lett. 26, 1516–1518 (2001).

13. Schaffer, C.B., Brodeur, A., Garcia, J.F., and Mazur, E., Micromachining bulk glass by use of femtosecond laser pulses with nanojoule energy, Opt. Lett. 26, 93–95 (2001).

14. Streltsov, A.M. and Borrelli N.F., Fabrication and analysis of a directional coupler written in glass by nanojoule femtosecond laser pulses, Opt. Lett. 26, 42–43 (2001).

15. Cerullo, G., Osellame, R., Taccheo, S., Marangoni, M., Polli, D., Ramponi, R., Laporta, P., and De Silvestri, S., Femtosecond micromachining of symmetric waveguides at 1.5 µm by astigmatic beam focusing, Opt. Lett. 27, 1938–1940 (2002).

16. Minoshima, K., Kowalevicz, A.M., Ippen, E.P., and Fujimoto, J.G., Fabrication of coupled mode photonic devices in glass by nonlinear femtosecond laser materials processing, Opt. Express. 10, 645–652 (2002).

17. Osellame, R., Taccheo, S., Cerullo, G., Marangoni, M., Polli, D., Ramponi, R., Laporta, P., and De Silvestri, S., Optical gain in Er-Yb doped waveguides fabricated by femtosecond laser pulses, Electron. Lett. 38, 964–965 (2002).

18. Schaffer, C.B., Jamison, A.O., Garcia, J.F., and Mazur, E., Structural changes induced in transparent materials with ultrashort laser pulses, in *Ultrafast lasers: Technology and applications*, edited by M.E. Ferman, A. Galvanauskas, and G. Sucha, Marcel Dekker, New York (2002), pp. 395–417.

19. Streltsov, A.M. and Borrelli, N.F., Study of femtosecond-laser-written waveguides in glasses, J. Opt. Soc. Am. B 19, 2496–2504 (2002).

20. Will, M., Nolte, S., Chichkov, B.N., and Tünnermann, A., Optical properties of waveguides fabricated in fused silica by femtosecond laser pulses, Appl. Opt. 41, 4360–4364 (2002).

21. Chan, J.W., Huser, T.R., Risbud, S.H., and Krol, D.M., Modification of the fused silica glass network associated with waveguide fabrication using femtosecond laser pulses, Appl. Phys. A 76, 367–372 (2003).

22. Nolte, S., Will, M., Burghoff, J., and Tünnermann, A., Femtosecond waveguide writing: a new avenue to three-dimensional integrated optics, Appl. Phys. A 77, 109–111 (2003).

23. Gorelik, T., Will, M., Nolte, S., Tünnermann, A., and Glatzel, U., Transmission electron microscopy studies of femtosecond laser induced modifications in quartz, Appl. Phys. A 76, 309–311 (2003).

24. Du, D., Liu, X., Korn, G., Squier, J., and Mourou, G., Laser-induced breakdown by impact ionization in SiO_2 with pulse widths from 7 ns to 150 fs, Appl. Phys. Lett. 64, 3071–3073 (1994).

25. Stuart, B.C., Feit, M.D., Herman, S., Rubenchik, A.M., Shore, B.W., and Perry, M.D., Optical ablation by high-power short-pulse lasers, J. Opt. Soc. Am. B 13, 459–468 (1996).

26. Lenzner, M., Krüger, J., Sartania, S., Cheng, Z., Spielmann, Ch., Mourou, G., Kautek, W., and Krausz, F., Femtosecond optical breakdown in dielectrics, Phys. Rev. Lett. 80, 4076–4079 (1998).

27. Oberson, P., Gisin, B., Huttner, B., and Gisin, N., Refracted near-field measurements of refractive index and geometry of silica-on-silicon integrated optical waveguides, Appl. Opt. 37, 7268–7272 (1998).

28. Conzone, S.D., Hayden, J.S., Funk, D.S., Roshko, A., and Veasey, D.L., Hybrid glass substrates for waveguide device manufacture, Opt. Lett. 26, 509–511 (2001).

29. Bansal, N.P. and Doremus, R.H., *Handbook of Glass Properties*, Academic Press, Orlando, FL (1986).

30. Harbsmeier, F. and Bolse, W., Ion beam induced amorphization in α quartz, J. Appl. Phys. 83, 4049–4054 (1998).

31. Innerhofer, E., Südmeyer, T., Brunner, F., Häring, R., Aschwanden, A., Paschotta, R., Hönninger, C., Kumkar, M., and Keller, U., 60-W average power in 810-fs pulses from a thin-disk Yb:YAG laser, Opt. Lett. 28, 367–369 (2003).

32. Limpert, J., Clausnitzer, T., Liem, A., Schreiber, T., Fuchs, H.-J., Zellmer, H., Kley, E.-B., and Tünnermann, A., High average power femtosecond fiber CPA system, Opt. Lett. 28, 1984–1986 (2003).

Integrated Planar Erbium-Doped Waveguide Amplifiers: Technological and Optical Considerations

Matthias Brinkmann[1,2], Steffen Reichel[1], and Joseph S. Hayden[3]

[1] SCHOTT Glas, P.O. Box 2480, Hattenbergstrasse 10, D-55014 Mainz, Germany
[2] Matthias Brinkmann is now with the University of Applied Sciences - Darmstadt, Mathematics and Science Faculty, Schoefferstrasse 3, D-64295 Darmstadt, Germany
brinkmann@fh-darmstadt.de
[3] SCHOTT Glass Technologies, Inc., 400 York Avenue, Duryea, PA 18643-2036

1 Introduction

Next-generation optical telecom devices need both novel optical materials and appropriate material modification technologies in order to realize all-optical functionalities like light generation, light guiding, signal splitting and combining, wavelength division (de-)multiplexing (WDM), amplification, gain equalization, switching, and detection [1]. In addition to glass fiber drawing techniques, methods for producing single-mode waveguides in planar devices have been moved into the focus of today's worldwide photonics research and development activities [2]. Especially in access networks, the significance of the above-listed signal-management operations is growing compared to the "pure" light transportation feature of optical fibers.

However, an established road map, predicting which planar technology will be used within the next years, is missing in today's optical telecom business. All competing materials and corresponding technologies like silica on silicon, $LiNbO_3$, InP, SiON, polymer, and glass have their advantages and disadvantages in realizing the above-mentioned optical functionalities in an integrated planar way [3]. However, any technology for optical integration has to face one severe problem: grooming losses. For many of the above-mentioned functionalities [i.e., (de-)multiplexing], signal splitting and combining, gain equalization, switching, and fiber–chip coupling and transmission losses between 1 and 10 dB occur. Integration means implementing many of these optical operations in one device and, hence, aligning the corresponding grooming losses in series. Therefore, compact amplifiers (EDWA: erbium-doped waveguide amplifier) with moderate gain (about 10 dB within the C-band), low power consumption, and the capability of integration are required for integrated devices.

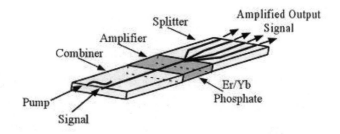

Fig. 1. Sketch of an integrated lossless splitter, consisting of a passive combiner, a planar waveguide amplifier, and a passive splitting section, all implemented on a hybrid substrate.

This chapter is organized as follows: In Section 2, a glass-based technological concept for the realization of integrated planar waveguide amplifiers is introduced and discussed. Section 3 deals with optical aspects of EDWAs and compares simulated with first experimental results.

2 Technological Concept of Glass-Based EDWAs

2.1 Hybrid Substrates

Because of the above-described lack of a lead material and technology for optical integration and the demand of "intermediate amplification," hybrid integration is one of today's most promising concepts for including all of the necessary optical signal processing functionalities into one single device [4]. For the most advanced method of hybrid integration, the optical bench should be one single optical chip [5]. As an example of this type of hybrid integrated devices, a lossless splitter is shown in Figure 1.

Here, the hybrid substrate consists of a combination of a active (e.g., Er/Yb-doped phosphate glass) and a passive glass. The single-mode waveguide system has to be realized by a burying technology, as described below. However, this concept of hybrid integration is not restricted to glass components.

2.2 Low-Temperature Bonding

For any kind of hybrid optical integration, joining and assembling technology play a very important role. In the following, a special inorganic joining technology, called low-temperature bonding from SCHOTT Glass, is described, as it applies to preparing a hybrid glass substrates for integrated photonic applications such as that depicted in Figure 1.

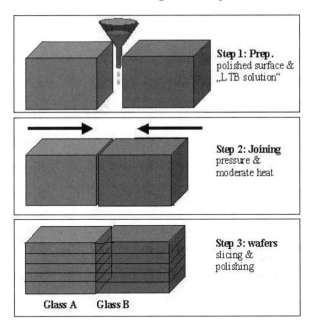

Fig. 2. Schematic diagram explaining the low-temperature bonding process as applied to mass production of hybrid substrates for integrated-optical photonic applications.

The low-temperature bonding process is shown schematically in Figure 2. Here, three steps are presented that, along with a structuring technology discussed later, form our hybrid integration concept for the fabrication of multiple numbers of identical integrated optical devices. Although low-temperature bonding can also be applied to structural applications, we emphasize in our discussion particular processing details that relate to fabrication of an optical quality interface between bonded parts.

Step 1 consists of preparing the surfaces that will eventually be bonded by standard grinding and polishing processes to a surface flatness of nominally 200 nm (peak to valley). These surfaces are then cleaned using a combination of a mild detergent, caustic solution, deionized H_2O, and volatile organics and then dried with deionized N_2.

The second step is best performed in a class 100 clean environment and consists of sandwiching a small volume of aqueous, inorganic adhesive between the prepared surfaces to create a hybrid preform. Bonding is achieved by condensation reactions that occur within the interfacial region to form a rigid bond at room temperature. Such rigid joints can be formed in time spans of as little as 15 s depending on the chosen chemistry and bonding conditions, but by proper selection of bonding solution, the time duration prior to setting of the bond can be adjusted to accomodate manual alignment or adjustments of part position. Bonded assemblies are optionaly further cured by heating at

Fig. 3. Photograph showing a hybrid preform and several hybrid substrates prepared from active Er/Yb-doped phosphate glass (grey) and passive phosphate glass (clear).

temperatures ranging from 60°C to 375°C. A typical hybrid preform formed by the low-temperature bonding process is shown in Figure 3.

A key feature of the low-temperature bonding technology is that robust bonds are formed at low temperatures and can be cut, ground, and polished using conventional fabrication technology into individual substrates for preparation of integrated-optical devices. The third step of our hybrid integration concept thus involves the sectioning of the hybrid preform by slicing to prepare parts that yield individual hybrid substrates. The importance of Step 3 is that it shows the applicability of low-temperature bonding to mass production, an important feature for any successful technology to be applied to next-generation optical telecom devices. Examples of such finished substrates are also shown in Figure 3.

It has been demonstrated that the strength of parts is nearly independent of the chosen heat treatment step [6] and both bonded silicate and phosphate glassy materials demonstrate average flexural strength values exceeding 35 MPa. The optical properties are also outstanding, with bond insertion losses less than 0.005 dB and reflection values from the bond interface of less than −34 dB [6].

2.3 Waveguide-Burying Technologies

The forth processing step in our hybrid integration is the burying of optical waveguides into the substrate. Coating technologies are obviously not suitable for any hybrid substrate concept. However, the huge advantage of this method is closely related to the order of the four processing steps: surface preparation, low-temperature bonding, slicing and polishing, and waveguide burying. Because the waveguides are introduced through the material interface *after* the joining process (assembly step), we avoid any type of time-consuming and cost-expensive aligning steps, which are normally necessary for other integration technologies. Notice that aligning and assembly are the dominant cost drivers for integrated telecom products. After evaluating all types of burying technologies, we have identified two promising candidates: ion exchange and laser structuring.

Ion Exchange

The principles and different methods of ion exchange for producing optical waveguides are already well explained in the literature [7]. Here, we only want to discuss the advantages and disadvantages for applying this technology to hybrid substrates. Ion exchange is already a well-established and commercially applied technology for optical telecom device manufacturing. The waveguide quality can be very high (low passive loss, excellent mode field match to standard telecom fibers, low polarization dependent loss (PDL)). However, the process parameters (e.g., choice of exchange partners, diffusion coefficients) strongly depend on the glass type. Hence, ion exchange can only be applied to hybrid substrates if the different materials (e.g., glasses) have very similar ion-exchange parameters (e.g., for the setup of Figure 1, if the only difference in the glass types is the addition of small amounts of Er/Yb for the active material).

Laser Structuring

By focusing a laser beam into a glass substrate and moving the focal point parallel to the surface, it is possible to induce a "line" of permanent increased refractive index into the material, which might act as a single-mode optical waveguide [8]. In principle, two different laser-structuring methods are currently under development. In ultraviolet (UV) laser structuring, photosensitive materials are required (e.g., Ge- or B-doped silicate glasses) [9]. The most recent method applies femtosecond-laser radiation (e.g., Ti:sapphire laser, 800 nm, 10 – 100 fs) to induce a refractive-index change in transparent materials via the physico-chemical reaction chain of multiphoton absorption, microplasma formation (optical breakdown), and material densification [8].

The sequential nature of the optical waveguide writing process by focused laser radiation has advantages and disadvantages for our concept of hybrid integration. Technologically, it is quite easy to change the laser-writing parameters (speed, intensity, parameters of the focusing optics) when writing across a material interface of the hybrid substrate. For each material, the optimal writing parameters can be automatically applied. However, because of the sequentially of the laser-writing process, the processing time is always higher than for any parallel structuring technique such as ion exchange or coating technologies.

An example for the success of our hybrid integration concept is displayed in Figure 4. Here, an optical waveguide has been buried into a hybrid phosphate glass. This result has been obtained using the equipment described in the chapter by Nolte. In the center of the image, the low-temperature bonding

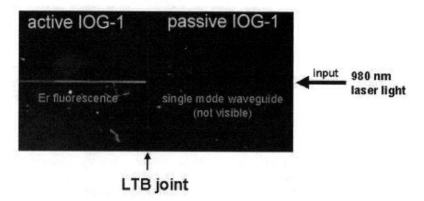

LTB joint

Fig. 4. Single-mode waveguide though an low-temperature bonding interface between active (i.e., Er doped) and passive phosphate glass IOG-1 (VIS microscope image).

interface between the active (left side) and the passive glass part is shown. The microscope image has been taken while sending 980-nm-pump light through the waveguide. In the active part, green fluorescence light can be seen along the waveguide, caused by the excited-state absorption of the pump light followed by radiative decay (green light radiation).

3 Optics of EDWAs

In addition to the development of technological and experimental facilities, appropriate theoretical and computational algorithms are important for predicting amplifier behavior and yielding improved device designs [10].

Today, user-friendly software tools for EDFA (erbium-doped fiber amplifier) design are already commercially available[4]. However, until now, there are only a few numerical algorithms known on the computation of amplification in EDWAs [11]. In contrast to fiber amplifiers, planar waveguide amplifiers show a nonrotational symmetric index profile and a different erbium distribution along the core and cladding region (as illustrated in Fig. 5). These features are significantly different from fiber amplifier setups and have to be taken into consideration for any realistic EDWA computation algorithm.

This section is organized as follows: First, the fundamental physics of planar waveguide amplifiers is implemented into the relevant amplifier equations. Then, the corresponding numerical procedure is compared to a commercial

---

[4] Among many tools, here are three selected examples:
   a. OASIX Amplifier Simulation Software, Lucent Technologies (1995)
   b. EDFA_Design, Optiwave Cooperation (1999)
   c. VPIcomponentMaker, VPI Systems (1999)

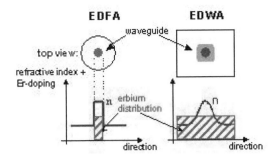

Fig. 5. Refractive-index and erbium concentration profile for a standard step-index EDFA and an EDWA.

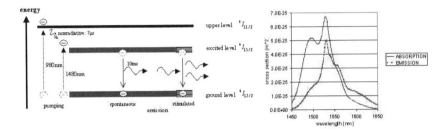

Fig. 6. Relevant energy levels of an erbium-doped amplifier [12] and typical cross sections.

code (see footnote on page 148, item b) for a standard EDFA. Afterward, results for EDWAs are compared with experimental data.

3.1 Theory

The relevant energy levels for an amplifying process in an erbium-doped amplifier can be seen in Figure 6. Two possible pumping wavelengths exist: 980 nm and 1480 nm. By pumping at 980 nm, erbium ions are excited from the ground state ($^4I_{15/2}$) to the upper level ($^4I_{11/2}$). From the upper state, the ions non-radiatively relax (7 μs [12]) to the excited lasing level. The excited level is metastable, and after about 10 ms, the ions fall back to the ground state. Typically, the signals are modulated much higher than $1/10\,\mathrm{ms} = 100\,\mathrm{Hz}$. Thus, the erbium ions cannot follow these fast oscillations and, hence, act as a low-pass filter (steady state) [10].

Due to the large difference in lifetimes (10 ms >> 7 μs), nearly all erbium ions are collected in the excited level, resulting in a *steady-state two-level system*, even when pumping at 980 nm [10]. Since the used glasses have a high dopability, upconversion processes can be neglected. The erbium-doped amplifier is completely described by the power propagation equation that is coupled to a corresponding rate equation. Spontaneous emission (acting as noise), stimulated emission, and also attenuation have to be included in a powerful algorithm for appropriate amplifier modeling.

The power propagation equation can be derived from the cross-section concept [13]:

$$
\frac{dI(x, y, z, f)dA}{dz} = [\sigma_{em}(f) \cdot N_2(x, y, z) -
$$
$$
-\sigma_{ab}(f)N_1(x, y, z)] \times I(x, y, z, f) \, dA +
$$
$$
+\sigma_{em}(f)N_2(x, y, z) \, dI_{\text{Noise}}(x, y, z, f) \, dA, \qquad (1)
$$

where I is the intensity of the light (W/m^2), σ_{em} (σ_{ab}) is the emission (absorption) cross section (m^2), N_1 (N_2) is the erbium ion density in the ground (excited) state $(1/\text{m}^3)$, dA is the incremental area perpendicular to the propagation direction, and dz is the incremental length along the propagation direction. The last term containing the incremental added intensity dI_{Noise} represents the spontaneous emitted noise.

Now, the following ansatz is used in order to separate $I(x, y, z)$ of Eq. (1), assuming the mode profile will not change during propagation:

$$
I(x, y, z) := \phi(x, y)P(z), \qquad (2)
$$

where $P(z)$ is the power (W) and $\phi(x, y)$ is the normalized mode field of the power $(1/\text{m}^2)$ obtained from the *transverse mode distribution by solving Maxwell's equations* for the specific geometry and refractive-index profile.

The normalized mode field has the following properties:

1. *Integration* is performed *only over the doped region*, because stimulated emission can only occur here.
2. $\int_{-\infty}^{+\infty} \int \phi(x, y) \, dx \, dy \equiv 1$.
3. $\int_{-\infty}^{+\infty} \int I(x, y, z) \, dx \, dy = P(z) \underbrace{\int_{-\infty}^{+\infty} \int \phi(x, y) \, dx \, dy}_{=1} \equiv P(z)$.

Note, ansatz (2) is also valid for the noise power, since noise is coupled into the guided mode. Ansatz (2) inserted into Eq. (1) followed by integrating over the whole transverse plane results in:

$$
\frac{dP}{dz} = \left[\sigma_{em}(f) \int_{-\infty}^{+\infty} \int N_2(x, y, z)\phi(x, y) \, dx \, dy -
$$
$$
-\sigma(f) \int_{-\infty}^{+\infty} \int N_1(x, y, z)\phi(x, y) \, dx \, dy \right] P(z) +
$$
$$
+\sigma_{em}(f) \, dP_{\text{Noise}}(z, f) \int_{-\infty}^{+\infty} \int N_2(x, y, z)\phi(x, y) \, dx \, dy, \qquad (3)
$$

where dP_{Noise} is the noise power emitted along the distance dz of the waveguide. A single photon with energy hf (Ws, i.e., W/Hz, where f is the frequency of the emitted photon and $h = 6.626 \times 10^{-34} \, \text{Ws}^2$ Planck's constant) contribute to the spectral noise power by $dP_{\text{Noise}}/df = hf$. Hence, the noise power can be calculated from

$$dP_{\text{Noise}} = hf\, df, \qquad (4)$$

where df is the resolution bandwidth chosen to resolve the noise power spectrum adequately. This noise power is coupled into all possible polarization modes m. Thus, the total noise power obeys

$$dP_{\text{Noise}} = mhf\, df, \qquad (5)$$

where $m = 2$ for the fundamental mode and $m = 4$ for the second-order mode [10]. Until now, additional background losses where not taken into account. These can easily be included by adding a term $-\alpha_{\text{bgl}}P(z)$ on the right-hand side of Eq. (3). For a detailed derivation, see Ref. [14]. Thus, the final *power propagation equation describing an erbium-doped optical amplifier* is given by (assuming the same background loss in core and cladding)

$$\frac{dP}{dz} = \left[\sigma_{em}(f) \int_{-\infty}^{+\infty} \int N_2(x,y,z)\phi(x,y)\, dx\, dy - \right.$$
$$\left. -\sigma(f) \int_{-\infty}^{+\infty} \int N_1(x,y,z)\phi(x,y)\, dx\, dy \right] P(z) + \sigma_{em}(f)$$
$$\cdot mh\, fdf \int_{-\infty}^{+\infty} \int N_2(x,y,z)\phi(x,y)\, dx\, dy - \alpha_{bgl}P(z). \qquad (6)$$

Note: The mode profile depends also on the frequency f and $P = P(z,f)$, but this is dropped for simplicity.

The erbium ion densities in different spectroscopic states are obtained from the rate equation (reducing to the two-level steady-state rate equation, as discussed). The following assumptions are made for the rate equations:
1. Neglecting the nonradiative decay lifetime, because it is much faster than the spontaneous emission lifetime [10].
2. Neglecting the upward and downward thermal stimulated terms in the rate equations (the optical frequency approximation) [13].
3. The power at a single wavelength is less than about $1\,\text{W}$ and thus the nonradiative decay rate (of the pump) is much larger than the pumping rate ($\to$ no "Q-switching effects") [10].

Under these well-confirmed assumptions [10], the rate equation (ions per second and unit volume) for the upper level is given by [15]:

$$\frac{dN_2}{dt} = 0 = W_{12}N_1 - W_{21}N_2 - \frac{N_2}{\tau_{\text{sp}}} \qquad (7)$$

where W_{12} (W_{21}) is the upward (downward) transition rate (ions per second). The spontaneous emission lifetime τ_{sp} is typically $10\,\text{ms}$ (see Fig. 2). In general, the transition rate can be determined by [9]

$$W_{12} = \frac{\sigma_{ab}(f)}{hf} I(x, y, z) \quad \text{and} \quad W_{21} = \frac{\sigma_{em}(f)}{hf} I(x, y, z), \tag{8}$$

where $I(x, y, z)$ denotes the intensity and f the carrier frequency of the incident wave. Using Eqs. (8) and (2), the rate equation for the upper level is converted into (with the lower level omitted for simplicity)

$$0 = \frac{\sigma_{ab}(f)}{hf} N_1(x, y, z)\phi(x, y)P(z) -$$

$$- \frac{\sigma_{em}(f)}{hf} N_2(x, y, z)\phi(x, y)P(z) - \frac{N_2(x, y, z)}{\tau_{\mathrm{sp}}}. \tag{9}$$

The rate equations for the upper and lower levels are related to each other by the *particle conservation*; that is,

$$N_1(x, y, z) + N_2(x, y, z) = N_{\mathrm{Erbium}}(x, y), \tag{10}$$

where N_{Erbium} is the erbium ion density. In almost all practical cases, the erbium ion density is constant (= spatial homogeneous distributed):

$$N_{\mathrm{Erbium}}(x, y) = \begin{cases} N_{\mathrm{Erbium}}^0, & \text{doped region} \\ 0, & \text{elsewhere.} \end{cases} \tag{11}$$

Suppose two signals are incident on the erbium-doped amplifier. The signals are characterized by their complex slowly varying envelopes $u_1(t)$ and $u_2(t)$, respectively. Hence, the total power P_{tot} is given by

$$P_{\mathrm{tot}} = |u_1(t)|^2 + |u_2(t)|^2 + 2|u_1(t)||u_2(t)|\cos(\Delta\omega t), \tag{12}$$

with $\Delta\omega = \omega_1 - \omega_2$ being the difference between the two carrier frequencies. Equation (12) describes power oscillating with time around the sum of the average power of each wave. Typically signals (as well as the spectral resolution) are separated by more than about $10\,\mathrm{kHz}$; thus, the power oscillation is too fast for the erbium ions and therefore canceled out. Hence, the power for the superposition of two waves is given by [10, 14],

$$P_{\mathrm{tot}} = P_1 + P_2. \tag{13}$$

For N waves, the rate equation is given by [10]

$$0 = \sum_{k=1}^{N} \frac{\sigma_{ab}(f_k)}{hf_k} N_1\phi_k(x, y)P_k(z)$$

$$- \sum_{k=1}^{N} \frac{\sigma_{em}(f_k)}{hf_k} N_2\phi_k(x, y)P_k(z) - \frac{N_2(x, y, z)}{\tau_{sp}}, \tag{14}$$

including pump, signal, and noise power. Note the transverse mode distribution $\phi(x, y)$ has also a subscript k representing a different frequency (the mode field changes with frequency).

Rearranging Eq. (14) with the help of Eq. (10) results in the final form of the rate equation:

$$N_2(x, y, z) =$$

$$N_{\text{Erbium}} \frac{\sum_{k=1}^{N} \frac{\sigma_{ab}(f_k)}{hf_k} \tau_{\text{sp}} \phi_k(x, y) P_k(z)}{1 + \sum_{k=1}^{N} [(\sigma_{ab}(f_k) + \sigma_{em}(f_k))/hf_k] \tau_{sp} \phi_k(x, y) P_k(z)}. \tag{15}$$

For an ideal two-level system, the absorption and emission cross section must be identical [13]. In Eq. (15), the cross sections are not identical, indicating that the *erbium-doped amplifier is not an ideal two-level system*, due to the Stark split of the upper and lower level into seven and eight sublevels (also drawn in Fig. 6).

The total light in the amplifier (i.e., the amplified spontaneous emission noise, the signal, and the pump light) is thought to be a number of N optically monochromatic waves, each of them having a carrier frequency f_k and being separated from adjacent waves by Δf. Thus, each frequency obeys the following propagation equation:

$$\frac{dP_k}{dz} = \left[\sigma_{em}(f_k) \int_{-\infty}^{+\infty} \int N_2(x, y, z) \phi_k(x, y) \, dx \, dy \right.$$

$$\left. - \sigma_{ab}(f_k) \int_{-\infty}^{+\infty} \int N_1(x, y, z) \phi_k(x, y) \, dx \, dy \right] P_k(z) + \sigma_{em}(f_k)$$

$$mhf_k \, df \times \int_{-\infty}^{+\infty} \int N_2(x, y, z) \phi_k(x, y) \, dx \, dy - \alpha_{\text{bgl}} P_k(z). \tag{16}$$

All N propagation equations are coupled by the rate equation (15) and particle conservation (9). In our computation algorithm, these equations are numerically solved in two steps. First, the mode distribution ϕ_k is numerically obtained from standard commercial beam-propagation software at several frequencies. Second, Eqs. (16), (15), and (9) are solved numerically using a fifth-order Runga–Kutta method with adaptive step size control for the differential equations (16). A finite-element procedure with rectangular elements of the Serendipity type (which is essentially the analogy of the trapezoidal rule in two dimensions) was implemented for the integrals in Eq. (16).

3.2 Application to EDFAs

In order to test the derived equations and its numerical implementation, a special case of an erbium-doped fiber amplifier was simulated (i.e., only the core is uniformly erbium doped in combination with a rotationally symmetric step-index profile). For this index distribution, an analytic solution of the mode profile exists and the propagation equation can be simplified due to the small erbium-doped core [10]. For a given setup our algorithm has been compared to a commercial software code (see footnote on page 148, item b)) (cross

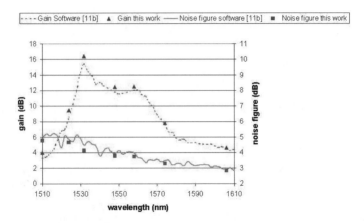

Fig. 7. Simulation results of an erbium-doped fiber amplifier using commercial software (see footnote on page 148, item b) compared with results from our algorithm (both using the same parameters), showing good agreement between both simulation tools.

sections are shown in Figure 6, core diameter: 3.5 μm, $N_{\text{Erbium}} = 1.23 \times 10^{26}$ 1/m^3, length: 18 cm, 45 mW to 980 nm forward pumped, numerical aperture: 0.264). Figure 7 compares the commercial software results with data obtained from our algorithm, demonstrating the validity of our model for the amplification process and the EDFA noise performance.

3.3 Application to EDWAs

After demonstrating the validity of our model for an EDFA, an EDWA was simulated. In contrast to a fiber amplifier, the whole glass block is homogeneously erbium doped for an EDWA. Its planar waveguide could, for instance, have been manufactured by masked ion exchange (see bottom inset of Figure 8).

For simplicity, the refractive-index profile was assumed to be Gaussian (see Fig. 8). From the refractive-index profile, the mode field at different wavelengths was calculated (see top inset of Fig. 8). The following parameters were used for the EDWA simulations: a Gaussian refractive-index profile with maximum $\Delta n = 0.02$, $\tau_{\text{sp}} = 10.1$ ms, $N_{\text{Erbium}} = 3.0 \times 10^{26}$ 1/m^3 (uniform over the whole glass), 100 mW forward pumping at 980 nm, a total of -20 dBm of signal power in the EDWA waveguide (spread over eight wavelengths at 1528, 1533, 1538, 1543, 1548, 1553, 1558, and 1563 nm), $\alpha_{\text{bgl}} = 0.1$ dB/cm, and the cross sections as shown in Figure 9.

The gain spectrum computed by our algorithm is shown in Figure 10. A maximum gain of 45 dB at about 1535 nm could be reached (at 10 cm length). Caused by the narrow cross-section peaks (see Fig. 9) of the active glass, only

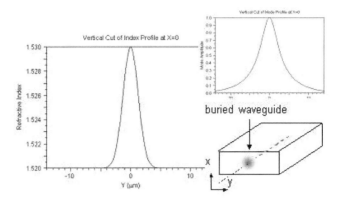

Fig. 8. Assumed Gaussian refractive-index distribution of a buried waveguide (full width half-maximum of 3 μm and Δn = 0.01). The top inset shows the numerically calculated (non-Gaussian) transverse mode field at 1550 nm. The bottom inset displays the 3-D geometry of a buried waveguide forming an EDWA.

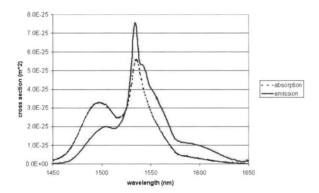

Fig. 9. Emission and absorption cross section of the erbium-doped glass.

limited broadband behavior of the EDWA is achieved, which makes this design unsuitable for a full C-band amplifier.

However, a proper refractive-index design (Δn = 0.01) together with adapted pump power and waveguide length could provide a much flatter gain spectrum. This is shown in Figure 11, where compares the normalized gain spectrum for the 18-cm waveguide with measurements on the actual device. Physically, this can be explained as follows: A smaller refractive-index difference Δn leads to significant broadening in the mode profiles for both the pump and all signal channels. Hence, this effect can be specifically used to simultaneously adjust the mode diameter ratios of all signal channels relative to the pump mode diameter (i.e., the radial erbium inversion distribution). This leads to a significant variation of the effective (i.e., radially averaged) erbium inversion level "seen" by the different signal channels. Thus, the index

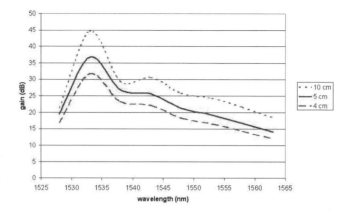

Fig. 10. Gain spectrum of an EDWA for various waveguide lengths.

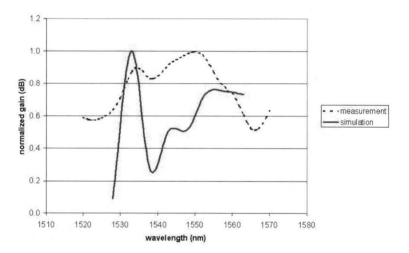

Fig. 11. Comparison of normalized gain spectrum obtained from simulation and measurement.

difference can be used as a design parameter to effectively broaden the EDWA gain spectrum.

Additionally, a measured gain spectrum displays a much more flattened behavior, as predicted by our simulations (see Fig. 11). The deviations between measurement and simulation could be explained by an incomplete knowledge of the various input data (needed for our computation algorithm). However, the measurement and the simulation show the same trend: A flatter gain spectrum can be achieved by a proper waveguide design!

4 Summary

Hybrid integration is one of today's most promising concepts in telecom networking for including all necessary optical signal processing functionalities into one single optical chip. Therefore, advanced material joining and waveguide structuring technologies have to be developed in order to provide low-loss waveguides and optical interfaces. Here, SCHOTT employs a technology called low-temperature bonding, which can be used to fabricate hybrid substrates from different transparent materials. In combination with our advanced optical glasses and suitable waveguide-burying technologies, these modification techniques could provide a powerful toolbox for the design of future integrated optical devices.

In addition, we successfully developed and implemented a rigorous algorithm for computing the optical properties of planar waveguide amplifiers. In contrast to most algorithms, our presented model can handle arbitrary refractive-index profiles and geometries together with an arbitrary erbium-dopant profile. This is of particular importance for designing planar waveguide amplifiers (EDWA). Thus, this algorithm enables detailed EDWA understanding and provides an effective design tool for complex integrated devices.

References

1. Ramaswami, R. and Sivarajan, K.N., *Optical Networks: A Practical Perspective*, Morgan Kaufmann, San Francisco, CL (1998).
2. Okamoto, K., *Fundamentals of Optical Waveguides*, Academic Press Orlando, FL (2000).
3. Yamada, Y., Takagi, A., Ogawa, I., Kawachi, M., and Kobayshi, M., Silica-based optical waveguide on terraced silicon substrate as hybrid integration platform, Electron. Lett. 29, 444 (1993).
4. CIR study, The market for integrated optical products: 2001-2005, Communications Industry Researchers, Inc., Charlottesville (Nov 2001)
5. Conzone, S.D., Hayden, J.S., Funk, D.S., Roshko, A., and Veasey, D.L., Hybrid glass substrates for waveguide device manufacture, Opt. Lett. 26, 509 (2001).
6. Hayden, J.S., Simpson, R.D., Conzone, S.D., Hickernell, R., Callicoatt, B., and Sanford, N.A., Passive and active characterization of hybrid glass substrates for telecommunication applications, SPIE Proc. 4645, 43 (2002).
7. Najafi, S. J., *Introduction to Glass Integrated Optics*, Artech House Publishers, Boston - London (1992).
8. Hirao, K., Mitsuyu, T., Si, J., and Qui, J., *Active Glass for Photonic Devices*, Springer Verlag New York (2001).
9. Lemaire, P. J., Atkins, R. M., Mizrahi, V., and Reed, W. A., High pressure hydrogen loading as a technique for achieving ultrahigh UV photosensitivity and thermal sensitivity in germania doped optical fibers, Electron. Lett. 29, 1191–1193 (1993).
10. Giles, R. and Desurvire, E., Modeling Erbium-Doped Fiber Amplifiers, IEEE J. Lightwave Technol. 9(2), 271–283 (1991).

11. Veasey, D.L., Gary, J.M., and Amin, J., Rigorous scalar modeling of Er and Er/Yb- doped waveguide lasers, SPIE Proc. 2996, 109–120 (1997). Huang, W., and Syms, R. R. A., Analysis of folded erbium-doped planar waveguide amplifiers by the method of lines, J. Lightwave Technol. 17(12), 2658-2664 (1999).

12. Desurvire, E. *Erbium-doped fiber amplifiers: Principles and Applications*, John Wiley & Sons, New York (1994).

13. Siegman, A., *Lasers*, University Science Books, Mill Valley, CA (1986).

14. Reichel, S. and Zengerle, R., Effects of nonlinear dispersion in EDFA's on optical communication systems, IEEE J. Lightwave Technol. 17(7), 1152–1157 (1999).

15. Saleh, B. and Teich, M., *Fundamentals of Photonics*, John Wiley & Sons, New York (1991).

Two-Dimensional Photonic Crystal Waveguides

Cecile Jamois[1], Ulrich Gösele[1], Ralf Boris Wehrspohn[2], Christian Hermann[3], Ortwin Hess[4], and Lucio Claudio Andreani[5]

[1] Max-Planck Institut für Mikrostrukturphysik, Weinberg 2, 06120 Halle, Germany,
jamois@mpi-halle.de,
[2] Nanophotonic Materials Group, Department of Physics, University of Paderborn, Paderborn, Germany
[3] DLR Stuttgart, Institut für Technische Physik, Theoretische Quantenelektronik, Pfaffenwaldring 38-40, 70569 Stuttgart, Germany
[4] Advanced Technology Institute, School of Electronics and Physical Sciences, University of Surrey, Guildford, Surrey, GU2 7XH, United Kingdom
[5] INFM and Dipartimento di Fisica A. Volta, Università di Pavia, Via Bassi, 6, 27100 Pavia, Italy

1 Introduction

In the last decades, a strong effort has been made to investigate and control the optical properties of materials, to confine light in specified areas, to prohibit its propagation, or to allow it to propagate only in certain directions and at certain frequencies. The introduction of components based on total internal reflection for light guidance, such as optical fibers or integrated ridge waveguides, has already been a revolution in the telecommunication and optical industry. In parallel to that, another way of controlling light based on Bragg diffraction has already been used in many devices like dielectric mirrors. In 1987, the principle of dielectric mirrors leading to one-dimensional light reflection was generalized to two and three dimensions [1,2], founding a new class of materials: photonic crystals. Since then, this new field has gained continuously increasing interest [3]. Photonic crystals (PCs) are materials with a periodic dielectric constant. If the wavelength of light incident on the crystal is of the same order of magnitude as the periodicity, the multiple-scattered waves at the dielectric interfaces interfere, leading to a band structure for photons. If the difference between the dielectric constants of the materials composing the photonic crystal is high enough, a photonic band gap (i.e., a forbidden frequency range in a certain direction for a certain polarization) can occur. However, a complete photonic band gap (i.e., a forbidden frequency range in all directions for all polarizations) can occur only in three-dimensional (3-D) photonic crystals. Although these 3-D photonic crystals look very promising and have been theoretically widely studied, their experimental fabrication is still a challenge [4–7]. Therefore, a strong effort has been invested to study two-dimensional (2-D) photonic crystals, which are much easier to fabricate and which still present most of the interesting properties of their 3-D counterparts. In the ideal case, 2-D photonic crystals are infinitely extended structures with a dielectric constant that is periodic in a plane and homogeneous in the third dimension. However, experimental structures are always finite, leading to scattering losses in the third dimension [8]. More recently, the concept of photonic crystal slabs consisting of a thin 2-D photonic crystal surrounded by a lower-index material has emerged and is now widely studied, because it offers a compromise between two and three dimensions. Indeed, combining the index guiding in the vertical direction with the presence of the photonic crystal in the plane of periodicity, a 3-D control of light can be achieved [9–11]. Among the several interesting effects in photonic crystals that can be used for a multitude of applications, such as modification of spontaneous emission [12,13] or effects based on the particular dispersion properties like birefringence [14], superprism effect, and negative refraction [15–17], one of the important effects relies on the existence of the band gap for waveguiding purposes. In this chapter, some properties of 2-D photonic crystals are studied, assuming first an infinite height (Section 2) and then a finite one (Section 3). Then, the influence of introducing a line defect into the photonic crystal lattice to build a waveguide is discussed, first in the case of infinite 2-D photonic crystals (Section 4) and finally in photonic crystal slabs (Section 5).

2 Infinite 2-D Photonic Crystals

Typically, 2-D photonic crystals consist of a lattice of parallel rods embedded in a substrate of different dielectric constant. This can be either air pores in a dielectric or dielectric rods in air ordered in a square or hexagonal lattice, such that the dielectric constant is homogeneous in the direction parallel to the rod axis – generally defined as the z direction – and periodic in the (x, y) plane:

$$\varepsilon(\mathbf{r}) = \varepsilon(\mathbf{r} + \mathbf{R}), \tag{1}$$

where $\mathbf{R}$ is any linear combination of the two unit vectors a_1 and a_2 of the 2-D photonic crystal lattice:

$$\mathbf{R} = l\mathbf{a_1} + m\mathbf{a_2}. \tag{2}$$

Due to the periodicity, the eigenfunctions of the system can be written in the form of Bloch states, in analogy to solid-state physics. In the case of the magnetic field,

$$\mathbf{H}_{n,\mathbf{k}}(\mathbf{r}) = e^{i\mathbf{k}\cdot\mathbf{r}} \cdot \mathbf{u}_{n,\mathbf{k}}(\mathbf{r}), \tag{3}$$

where n is the frequency band index, $\mathbf{k}$ is the wave vector, and the function $\mathbf{u}_{n,\mathbf{k}}$ has the periodicity of the photonic crystal:

$$\mathbf{u}_{n,\mathbf{k}}(\mathbf{r}) = \mathbf{u}_{n,\mathbf{k}}(\mathbf{r} + \mathbf{R}) \tag{4}$$

If the materials building the photonic crystal are assumed to be linear, isotropic, nonmagnetic, and free of charges, the following wave equation is obtained by combining Maxwell's equations:

$$\nabla \times \left(\frac{1}{\varepsilon(\mathbf{r})} \nabla \times \mathbf{H}(\mathbf{r}) \right) = \frac{\omega^2}{c^2} \mathbf{H}(\mathbf{r}) \tag{5}$$

This is an eigenvalue problem where the eigenvectors $\mathbf{H}(\mathbf{r})$ are called harmonic modes, and the eigenvalues $(\omega/c)^2$ are proportional to the squared frequency of these modes, c being the speed of light.

By solving the master equation (5) for k-vectors along the irreducible Brillouin zone of the photonic crystal, the band structure of the photonic crystal is obtained. Because the (x, y) plane of periodicity of the 2-D photonic crystal is a mirror plane of the system, the polarizations decouple; that is, the modes can be separated into transverse electric (TE) modes having only H_z, E_x, and E_y as non zero components, and transverse magnetic (TM) modes with the only non zero components E_z, H_x, and H_y. Since TE (resp. TM) modes have their magnetic (resp. electric) field oriented along the pore axis, they are sometimes also called H (resp. E) modes. The band structures for TE and TM polarizations are usually completely different, because the electric field (resp. magnetic field) for TE and TM polarizations is oriented in different directions relatively to the dielectric interfaces within the photonic crystal. In particular, band gaps can exist for one polarization and not for the other, or the position of the band gaps can be very different. It has turned out that for systems consisting of dielectric cylinders in air a complete band gap (i.e., a band gap for both polarizations) can be obtained only in a honeycomb lattice, whereas the hexagonal lattice of air holes in dielectric opens up a complete 2-D photonic band gap for a dielectric contrast n_2/n_1 larger than 2.6 [3,18].

Figure 1 gives an example of a band structure in the case of a hexagonal lattice of air pores in silicon with a relative radius r/a of 0.43, where a is the lattice constant of the photonic crystal. The band structure calculation was performed using the MIT package, a block-iterative frequency-domain code [19] with a grid of 64 points per lattice constant, yielding good convergence of the results. For this relative radius value, a large TE band gap exists from 0.275 to 0.460 in normalized frequency $\omega a/2\pi c$ and a smaller TM band gap from 0.385 to 0.405 that overlaps with the TE band gap, leading to a complete 2-D band gap in this frequency range. The variation of the photonic band-gap position with relative pore radius (so-called gap map) for a hexagonal lattice of air pores in silicon is shown in Figure 2. In this system, the complete band gap exists only for a relative radius larger than 0.4, the largest gap–midgap ratio – ratio between band gap width and midgap frequency – being 16.3% for a pore radius $r/a = 0.478$. However, such very large relative radius values are quite difficult to achieve experimentally. Therefore, most of the work based on the existence of a band gap in 2-D photonic crystals has focused on the TE

162   Cecile Jamois et al.

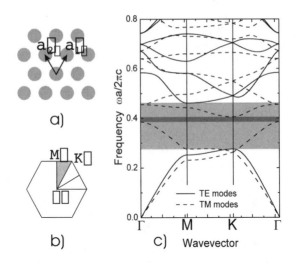

Fig. 1. (a) A 2-D hexagonal lattice and (b) its first Brillouin zone with the irreducible Brillouin zone delimited by the three high-symmetry points Γ, M, and K. (c) Band structure for a hexagonal array of air pores in silicon ($\varepsilon = 11.6$, $r/a = 0.43$) along the k-path Γ-M-K-Γ for TE (solid lines) and TM (dashed lines) polarizations. The light a region highlights the TE band gap and the dark a region highlights the TM (resp. complete) band gap (MIT package calculation).

band gap only, which is still quite large for smaller radii (e.g., at $r/a = 0.366$, the gap–midgap ratio for TE modes is as large as 42.5%).

From an experimental point of view, the approximation of infinitely long pores or rods can be done only in structures exhibiting high aspect ratios (ratio between pore/rod length to pore/rod diameter). Three-dimensional numerical simulations show that a 2-D photonic crystal can be approximated only by structures with aspect ratios larger than 20 [20]. This is difficult to achieve with conventional dry-etching techniques. However, a good candidate for the experimental study of 2-D photonic crystals is macroporous silicon, consisting of a periodic array of air pores in silicon. Indeed, in these structures prepared by photo-electrochemical dissolution of silicon in hydrofluoric acid [21, 22], very high aspect ratios up to 500 can be obtained, as illustrated in Figure 3. It has been recently shown that the optical properties of macroporous silicon can be well described by 2-D simulations [23, 24].

Figure 4a shows the reflectivity of TE-polarized light incident on a photonic crystal made of macroporous silicon in the Γ-M direction. The photonic crystal has a hexagonal lattice of air pores with a relative radius r/a of 0.366 and a lattice constant of 700 nm. The reflection was measured by using a Fourier transformation infrared (FTIR) spectrometer with attached optical IR microscope and calcium fluoride beam splitter covering a spectral region

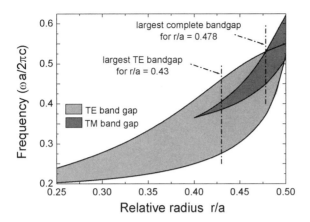

Fig. 2. A 2-D gap map (normalized frequency $\omega a/2\pi c$ versus relative pore radius r/a) for a hexagonal lattice of air holes in silicon ($\varepsilon = 11.6$). The position of the largest gap–midgap ratio is indicated (MIT package calculation).

Fig. 3. SEM image of a two-dimensional hexagonal lattice of air pores in silicon with a lattice constant of 1.5 μm. The pore depth is around 100 μm. The beveled etched part in front reveals the high uniformity of the structure from the top down to the bottom of the pores. (Courtesy of A. Birner, MPI Halle.)

from ultraviolet (UV) to mid-IR. The light source was a broadband tungsten lamp. The band structure of the photonic crystal is presented in Figure 4b. However, these curves cannot be compared directly with the band structure, because not all of the bands yield very high transmission. Indeed, for the main directions Γ-M and Γ-K, the plane defined by the wave vector and the pore axis is a mirror plane of the crystal. Thus, the modes can be separated into

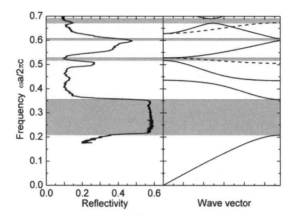

Fig. 4. (a) Reflectivity of TE-polarized light incident onto a 2-D photonic crystal made of a hexagonal lattice of air pores in silicon with a relative radius r/a of 0.366 and a lattice constant of 700 nm. The light beam is parallel to the plane of periodicity of the crystal in the Γ-M direction. (Courtesy of S. Schweizer (sample) and S. Richter (measurement), MPI Halle). (b) Corresponding band structure for TE polarization (MIT package calculation). The modes are sorted as laterally even (dashed lines) or laterally odd (solid lines) modes. The regions where no odd mode exists are highlighted in a and correspond to the high-reflectivity regions.

laterally even modes or odd modes. Since the $E_\parallel$- or $H_\perp$-field components always have to be considered to define the mode symmetry, laterally even (odd) symmetry corresponds to laterally odd (even) H_z-field distribution. The lateral symmetry of the modes is illustrated in Figure 5, taking as examples the H_z-field distribution of the first and fourth bands at the M-point. While coupling from an incident plane wave into the photonic crystal, only the odd modes are excited. Therefore, the zero-transmission (resp. high reflectivity) regions do not correspond only to band gaps, but to frequency regions where no laterally odd mode exists. Therefore, the agreement between the measured transmission and the calculated band structure is very good.

3 Photonic Crystals Slabs

Since 2-D photonic crystals cannot, by definition, provide light confinement in the direction parallel to the pore axis, a way to avoid out-of-plane losses is the use of photonic crystal slabs. Slab structures consist of a thin 2-D photonic crystal (core) surrounded by two layers of lower effective refractive index (claddings) that provide an index guiding by total internal reflection in the direction normal to the plane of the crystal, as in a planar waveguide. Different examples of photonic crystal slabs are shown in Figure 6. The first one,

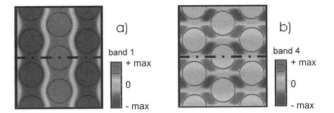

Fig. 5. H_z-field distribution of the (a) first and (b) fourth TE bands at the M-point for a 2-D photonic crystal consisting of a hexagonal array of air pores in silicon with $r/a = 0.366$ (MIT package calculation). If we consider the mirror plane defined by the Γ-M direction and the direction of the pore axis (dashed line), the first band has an even H_z-field distribution (corresponding to odd symmetry) and the fourth band an odd one (corresponding to even symmetry).

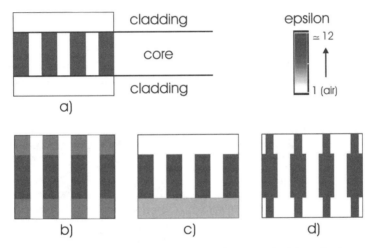

Fig. 6. Four examples of photonic crystal slabs (side view): (a) air-bridge structure with high index contrast between core and cladding (e.g., Si membrane); (b) low-index contrast heterostructure (e.g., AlGaAs/GaAs/AlGaAs system); (c) asymmetric structure with two different claddings, the upper one usually being air, current in SOI-based photonic crystal slabs; (d) modulated-pores structure achieved (e.g., in macroporous silicon).

the so-called "air-bridge" structure has been already largely studied [25,26]. It consists of a thin 2-D photonic crystal in a high-index membrane surrounded by air (Fig. 6a). In this case, the index contrast between the core and cladding is very high and light is strongly confined within the core. The second example (Fig. 6b) is a photonic crystal slab made in a heterostructure, usually consisting of III–V semiconductor materials. In this case, the index contrast between core and cladding is low; thus, the mode profiles are quite extended and the pores penetrates deeply into the cladding layers [27, 28]. The third type of structure (Fig. 6c) is a hybrid case often found in silicon on insulator (SOI) based systems [26,29,30]. The lower cladding usually consists of an oxide layer, not necessarily structured, whereas in most cases the upper cladding consists only of air, which leads to an asymmetric structure. The first three examples have in common that the effective index contrast between core and cladding is obtained by taking another material, as in the slab, to build the claddings. In the last case (Fig. 6d), the effective index contrast is obtained by modulation of the pore diameter, keeping the same material. In the cladding, the air-filling fraction is higher, thus the effective refractive index is lower. This kind of structure can be obtained for example in macroporous silicon [20].

Because of the finite height of the structure, polarization mixing occurs and the modes are no longer purely TE (resp. TM) polarized. On the other hand, if the (x, y) plane in the middle of the slab is a mirror plane of the structure (i.e., if both claddings are identical as in the examples shown in Figures 6a, 6b and 6d), the modes can be separated into vertically even (with the H_z component having a symmetric field distribution) and odd modes (antisymmetric field distribution). However, the first-order modes (i.e., modes having no node in the vertical direction) have field distributions within the core that are very similar to the corresponding modes existing in infinite 2-D photonic crystals. Furthermore, in the (x, y)-mirror plane itself, these modes are purely TE (resp. TM) polarized [31]. Thus, for first-order modes, the polarization mixing is quite small and the approximation even $\sim$ TE polarized and odd $\sim$ TM polarized can be assumed. Therefore, the terminologies TE (or H) modes and TM (or E) modes are found very often to refer to even and odd modes, respectively.

In addition to the polarization mixing effect, another important difference between infinite 2-D photonic crystals and photonic crystal slabs lies in the role of the light line, which is the lowest band in the cladding system. Indeed, the structure supports two kinds of mode (Fig. 7b). If the slab thickness is not too small and the index contrast between core and cladding is not too low, there exist some states in the slab below the light line. Since for this (k, ω) set no state is allowed in the claddings, the modes of the slab are totally internally reflected at the interface between core and claddings and confined within the core. These modes are pure Bloch modes; they are lossless in the case of an ideal structure where no scattering occurs. Above the light line of the cladding material, the modes lie within the continuum of leaky modes of the planar waveguide. Therefore, they are resonant or "quasiguided," because they have intrinsic radiation losses related to out-of-plane diffraction. Their lifetime is varying; it can be very long (weak radiation losses) as well as very limited (strong radiation losses). The larger the index contrast between core and claddings, the more modes exist below the light line [9,32]. For the type of heterostructure presented in Figure 6b, it is even usual that no guided mode exists at all. A method to estimate a posteriori the radiation losses consists in including an effective loss into a 2-D model through a dissipation mechanism (i.e., by inserting an imaginary index in the air holes and calculating the

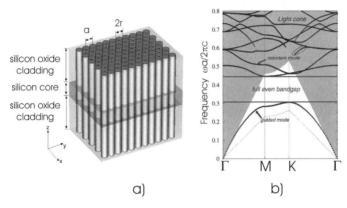

a)   b)

Fig. 7. (a) Photonic crystal slab consisting of a silicon core and two structured silicon oxide claddings (IOSOI structure) with air pores arranged in a hexagonal lattice. (b) Band structure of such a system having a relative thickness $h/a = 0.4$ and a relative radius of 0.366, even modes only. Above the light line of the cladding system (a region), the modes are resonant; their lifetime can be very limited. Only below the light line (white region), the modes are guided and lossless. The position of the lowest second-order mode is highlighted by a circle (2-D plane-wave calculation taking as basis the eigenmodes of the corresponding planar waveguide).

radiation losses using a first-order perturbation approximation). This very efficient method has been first developed for a transfer-matrix code [8] and then extended to a time-domain method [33].

Much theoretical work on photonic crystal slabs has focused on guided modes only [9, 10]. However, for practical applications, the modes above the light line must be considered too, because an incident wave can couple light to all of the modes existing at a given frequency in a given direction, provided they have the proper symmetry. Due to the finite height of the structure, 3-D calculations are necessary to calculate the band structure. However, several 3-D codes – like the MIT package – assume a periodicity in all three dimensions, which is not convenient for a photonic crystal slab. Due to the fictive periodicity in the vertical direction, some additional coupling between resonant modes occurs, which disturbs completely the band structure. Therefore, only guided modes can be calculated correctly, because they do not feel the fictive vertical periodicity. There are different methods to calculate resonant modes. One way is to use a 3-D finite-difference time-domain (FDTD) code with open-boundary conditions on the top and the bottom of the structure [34]. It is also possible to perform 2-D plane-wave calculations taking as a basis the eigenmodes of a planar waveguide where each layer has the same effective refractive index as the photonic crystal slab structure [32]. This is the method used in Figure 7.

If the slab thickness is increased, the cutoff frequency of the higher-order modes decreases, exactly as in a planar waveguide. For too thick slabs, higher-

168   Cecile Jamois et al.

order modes can exist within the first-order band gap of the photonic crystal slab. Thus, if these modes happen to have the right symmetry properties to be excited by an external light beam, they can limit the band gap or even destroy it completely. Reducing the vertical index contrast Δn, the cutoff frequency of the higher-order modes increases. A first guess to determine the cutoff frequency of the lowest second-order mode is to use the planar waveguide approximation. In this very simple approximation, we calculate the cutoff frequency of the second mode in a planar waveguide where each layer has the same effective refractive index as the photonic crystal slab, using the following relation:

$$\frac{h}{\lambda_0} = \frac{1}{2.\sqrt{n_2^2 - n_1^2}},$$ (6)

where h is the thickness of the core, λ_0 is the cutoff wavelength, n_1 and n_2 are the effective indices in the claddings and in the core, respectively, and for a photonic crystal,

$$\frac{h}{\lambda_0} = \frac{h}{a} \frac{a}{\lambda_0},$$ (7)

with h/a the relative thickness of the slab and $a/\lambda_0 = \omega a/2\pi c$ the normalized cutoff frequency of the mode. There are different methods to determine the effective refractive index of a photonic crystal. We choose here to consider the light lines of the 2-D systems corresponding to core and claddings and take the inverse of their tangent at the Γ-point. For example, for a photonic crystal slab consisting of a silicon slab ($\varepsilon = 11.6$) and two structured silicon oxide claddings ($\varepsilon = 2.1$) with a relative radius of 0.366, the effective refractive indices are 2.57 in the core and 1.25 in the claddings. For a relative silicon thickness of 0.4, this leads to a cutoff frequency of 0.56 for the first second-order mode. Such a system as well as the corresponding band structure have already been presented in Figure 7. It can be seen on the band structure that the lowest second-order mode has a cutoff frequency around 0.57, which is very close to the value calculated using the planar waveguide approximation. Therefore, this very simple method gives already a good guess of the cutoff frequency of higher-order modes in photonic crystal slabs.

4 Waveguides in Infinite 2-D Photonic Crystals

If a line defect is introduced into the 2-D photonic crystal lattice (e.g., by changing the pore radius of an entire pore line or removing it completely), some defect states are created. For a convenient design of the defect, some of these states ought to be located within the band gap of the photonic crystal. Since light cannot propagate in the photonic crystal at this frequency, it is localized in the surrounding of the defect in this case (i.e., the line defect acts as a waveguide). Figure 8 shows some examples of linear waveguides in 2-D photonic crystals.

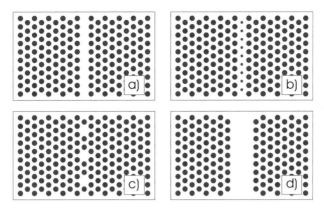

Fig. 8. Some examples of waveguides in 2-D photonic crystals: (a) W1 waveguide (i.e., waveguide having a width of one pore row) consisting of a row of missing pores, (b) W1 waveguide consisting of a row of pores with smaller diameter, (c) coupled-cavity waveguide, and (d) W3 waveguide (i.e., three pore rows wide).

The introduction of a line defect induces a symmetry breaking. Indeed, now the translation symmetry exists only in the direction parallel to the defect. Therefore, the new Brillouin zone is one dimensional, and the band structure of the 2-D photonic crystal has to be projected onto the k-path $\Gamma(0)$–$J(\pi/a)$ of the new Brillouin zone [35]. Figure 9 shows the band structure for TE modes of a W1 waveguide consisting of a row of missing pores in the Γ-K direction, the photonic crystal being made of a hexagonal lattice of air pores in silicon with relative radius 0.43. The band structure for the corresponding bulk photonic crystal has already been presented in Figure 1. The a regions in Figure 9 correspond to the continuum of projected bands of the bulk photonic crystal. Comparison with Figure 1 shows that there is a TE band gap in the frequency region from 0.275 to 0.460, where several defect states are located. However, not all of these defect states are guided due to the presence of the photonic band gap. Indeed, in such a structure, two guiding mechanisms coexist. The first one is based on the existence of the photonic band gap and the second one is classical index guiding due to the effective index contrast between the waveguide and its surroundings.

Since their existence is based on index contrast, index-guide modes (dashed lines) exist below the first band of the bulk photonic crystal. When they reach the J-point at the limit of the first Brillouin zone, they are folded back into the first Brillouin zone and continue to increase linearly in direction of the Γ-point, and so on. Figure 10 shows a comparison between the field distributions at the J-point of the two index guided modes of Figure 9 and those of the corresponding 2 D ridge waveguide having the same width and the same effective indices in core and claddings as the photonic crystal waveguide. There is good agreement between the fields of both systems, taking into account that the index-guided modes inside the photonic crystal wave-

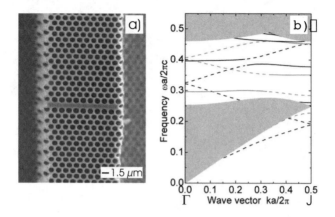

Fig. 9. (a) SEM picture and (b) band structure for the TE polarization of a linear waveguide in a 2-D photonic crystal. The photonic crystal consists of a hexagonal lattice of air pores in silicon with lattice constant 1.5 μm and relative radius 0.43. The linear defect is made of a row of missing pores in the Γ-K direction. The defect modes are sorted into index guided (dashed lines) and photonic-band-gap guided modes (solid lines) as well as into laterally even (a lines) and odd modes (black lines) (MIT package calculation).

guide are perturbed by the periodicity of the pores, which leads to the stop gap between the modes at the J-point. Considering first the two lowest-order modes (having at the J-point the frequencies 0.190 and 0.193, respectively), it can be noticed that the first index-guided mode in the W1 waveguide extends more into the photonic crystal than the second one. This can be explained by the fact that the wavelength of the lowest mode is larger (the mode extension for fundamental modes is $\lambda/2$). For the third and fourth bands, having at the J-point the frequencies 0.228 and 0.285, respectively, and corresponding to the second-order mode in the ridge waveguide, the same phenomenon occurs, but it is much stronger. Indeed, even if the wavelength is smaller, now the mode extension is λ, so that the third mode extends significantly into the surrounding photonic crystal. Thus, its field intensity is located partially in the air pores. Since the mode extension of the fourth mode is again smaller, this mode is very well located in the dielectric. Due to the important difference between the field distributions of these second-order modes, a much larger energy gap is expected between them than between the first-order ones. As a consequence, the fourth band lies within the band gap of the photonic crystal.

Unlike the index-guided modes, the photonic-band-gap-guided modes exist only within the band gap of the photonic crystal (solid lines in Fig. 9). Their guiding mechanism is based on the absence of allowed states in the surrounding photonic crystal and therefore presents a metalliclike behavior. Figure 11 shows the field distributions of the three lowest photonic-band-gap-

Ridge waveguide   Photonic crystal W1

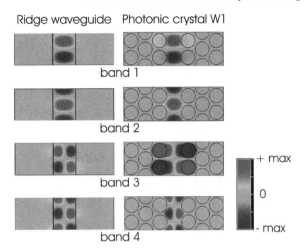

band 1

band 2

band 3

band 4

+ max

0

- max

Fig. 10. Comparison between the H_z-field distributions of the four lowest index-guided modes in the photonic crystal waveguide presented in Figure 9 (right) and in the corresponding ridge waveguide (left) having the same width and the same effective indices in the core (3.4 for silicon) and the claddings (1.55 for the system silicon/air pores with $r/a = 0.43$). The black lines on the left demarcate the silicon core of the ridge waveguide and the black circles on the right indicate the position of the air pores of the photonic crystal (MIT package calculation).

guided defect modes at the J-point, having frequencies 0.262, 0.352, and 0.378, respectively.

Another effect that can be noticed in Figure 9 is anticrossing. Indeed, only modes having different symmetries can cross each other without being disturbed. Modes having the same symmetry interact, leading to anticrossing effects, as can be seen for the two laterally odd modes in the middle of the gap at a frequency around 0.39. The interaction between these two modes illustrates the fact that the distinction made between index and band-gap guidance is not always a rigorous one.

Figure 12 shows the transmission through a W1 waveguide made of one row of missing pores in macroporous silicon, similar to that presented in Figure 9. The line defect is 27 μm long (18 lattice constants). The transmission through the waveguide was measured with a pulsed-laser source having a bandwidth of 200 nm and tunable over a large frequency range of the TE band gap (3.1 < λ < 5.5 μm) [24]. The measured spectrum in Figure 12a exhibits pronounced Fabry–Perot resonances over a large spectral range which are caused by multiple reflections at the waveguide facets and is in very good agreement with the corresponding 2-D FDTD transmission calculation of Figure 12b. Again, as in the case of bulk photonic crystals, only the waveguide modes with laterally odd symmetry with respect to the mirror plane in the middle of the waveguide can be excited by the laser beam, so that the even modes do not contribute to the transmission.

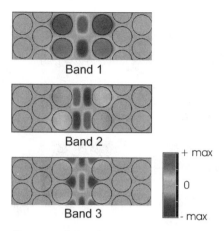

Band 1

Band 2

Band 3

+ max

0

- max

Fig. 11. H_z-field distributions of the three lowest photonic-band-gap-guided defect modes of the W1 waveguide presented in Figure 9. The black circles indicate the position of the air pores of the photonic crystal (MIT package calculation).

Due to the photo-electrochemical fabrication process, the diameter of the pores in the adjacent rows to the waveguide is increased, as can be seen in Figure 9a. This leads to a shift of defect modes to higher frequencies. In the transmission calculation, this particularity has been taken into account. Therefore, if the reflection curves are compared with the band structure of Figure 9, the small stopgap observed in the transmission curves for frequencies around 0.45 corresponds to the anti crossing between odd modes in the middle of the band gap.

5 Waveguides in Photonic Crystal Slabs

If a line defect is introduced into a photonic crystal slab lattice, the same phenomenon occurs as in the case of infinite 2-D photonic crystals. For an appropriate design of the waveguide, some defect states are located within the band gap of the photonic crystal, so that light is confined within the plane of the crystal in this frequency range. Combining this in-plane confinement with the vertical confinement due to the index contrast in the vertical direction, a 3-D light confinement is possible within a waveguide in a photonic crystal slab [25–29, 35]. Figure 13 shows the band structure of a W1 waveguide made of one row of missing pores in a photonic crystal slab, calculated using the MIT package. The photonic crystal is made in a silicon core with relative thickness 0.4 surrounded by two structured silicon oxide claddings (i.e., the air pores extend into the oxide claddings). The pores are arranged in a hexagonal lattice and have a relative radius of 0.366. The corresponding band structure for the bulk photonic crystal has been shown in Figure 7. The lower a region corresponds to the continuum of projected bulk bands, and the upper a region

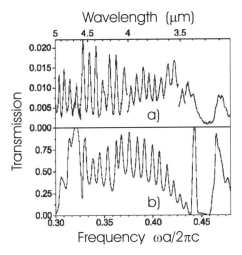

Fig. 12. (a) Transmission measurement and (b) FDTD calculation for the photonic crystal waveguide presented in Figure 9. The waveguide is 18 lattice constants long. (Reprinted from Ref. [36], © OSA.)

corresponds to the light cone. The defect modes below the region of projected bulk modes are index guided, as in the case of an infinite 2-D photonic crystal. Furthermore, the field distributions (1, 2 and 3) in the (x, y)-mirror plane in the middle of the silicon slab shown in Figure 13 are very similar to the corresponding ones presented in Figure 10. This is related to the fact that the defect modes are very well confined within the silicon core, as can be seen in the vertical cross sections shown in Figure 13. Above the projected bulk modes, the defect modes lie within the band gap of the photonic crystal. They are vertically confined as long as they are in the white region below the light line. For in-plane confinement, both guiding mechanism coexist as in the case of infinite 2-D waveguide structures: either the modes are index guided (like band 5) or they are guided due to the existence of the photonic band gap (like band 4). Again, comparison between the field distributions of these two defect modes and the two corresponding ones presented in Figures 9a and 10 (band 4) shows very strong similarities.

Above the light line, the defect modes become resonant; that is, they are still guided in the plane along the line defect but they are lossy in the vertical direction. Due to the intrinsic radiation losses, the light transmission through waveguides in a photonic crystal slab based on defect modes above the light line can be quite low. Figure 14a shows the band structure of a photonic crystal waveguide having a relative silicon thickness of 0.3, all the other parameters being the same as in Figure 13. The corresponding transmission through this waveguide with 30a (resp. 40a) length is presented in Figure 14b. The calculation takes into account light in- and out-coupling to the photonic crystal waveguide through a ridge waveguide having the same width and the

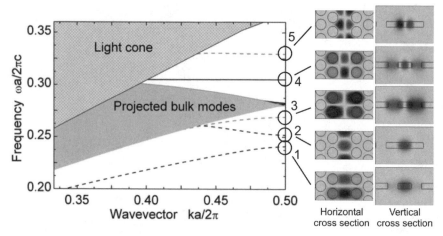

Fig. 13. (Left) Band structure of a W1 waveguide made of one row of missing pores in a photonic crystal slab. The photonic crystal is made in a silicon core with relative thickness 0.4 surrounded by two structured silicon oxide claddings (MIT package calculation). The pores are arranged in a hexagonal lattice and have a relative radius of 0.366. The lower a region corresponds to the continuum of projected bulk bands, and the upper a region correspondes to the light cone. The defect modes are sorted into index-guided (dashed lines) and photonic-band-gap-guided modes (solid lines) as well as into laterally even (a lines) and odd modes (black lines). **(Right)** H_z-field distributions of the five defect modes which are guided at the boundary of the first Brillouin zone (J-point). The horizontal cross sections show the field distributions in the (x, y)-mirror plane in the middle of the silicon slab, the black circles indicating the position of the pores. The vertical cross sections show the field distributions in the (x, z) planes containing the intensity maxima, the black lines indicating the position of the silicon slab.

same effective indices, leading to coupling losses of around 50%. The relative frequency range shown (0.32–0.44) corresponds to the photonic band gap of the crystal. Since odd modes are of interest, only the two odd defect bands are shown in the band structure, separated by a mini stop gap around 0.405. The light line of the cladding system is shown (black line). Thus, the first defect mode is guided only in the wave vector range 0.4–0.5, which corresponds to a very small frequency range. The extreme flatness of the band under the light line, corresponding to very small group velocity, is, in most cases, not desirable because it leads to very low transmission. One way to increase the group velocity as well as the transmission window is to vary the defect width [29]. The main part of the lower defect band in Figure 14 (for wave vectors below 0.4), contributing to the transmission, is resonant. From the difference in transmission between the two waveguide lengths, the attenuation due to radiation losses is estimated to be in the order of 100 dB/mm for a lattice constant of 500 nm. This estimation is in good agreement with theoretical predictions [37]

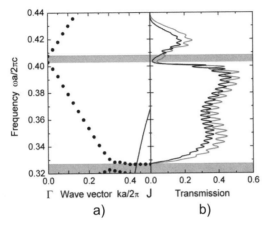

Fig. 14. (a) Band structure for odd modes of a photonic crystal waveguide having a relative silicon thickness of 0.3 and all other parameters as in Figure 13 (3-D FDTD calculation with open-boundary conditions on the top and the bottom of the structure). Only the frequency range between 0.32 and 0.44 is shown, corresponding to the photonic band gap of the crystal. The a regions correspond to the frequency regions within the photonic band gap, where no defect mode exists. (b) Transmission through the waveguide with 30 (a line) and 40 (black line) lattice constants length, respectively. The light is coupled in and out through a ridge waveguide, having the same width and the same effective indices as the photonic crystal waveguide.

as well as experimental measurements on W1 waveguides in silicon [38]. The upper odd defect mode is entirely resonant and the corresponding transmission is very small, indicating strong radiation losses ($\geq 300\,\mathrm{dB/mm}$). Losses for modes above the light line in a slab waveguide are expected to be proportional to the square of the vertical index contrast ($\Delta\varepsilon^2$) [8]. Thus, in III–V-based heterostructures, losses are much lower than in SOI-based waveguides. However, in silicon-based structures, some modes exist below the light line, being guided and therefore having very small intrinsic losses.

6 Conclusion

In summary, 2-D photonic crystals offer great possibility to guide light in two and three dimensions, by combining two different guiding mechanisms: index guiding and photonic-band-gap guiding. A line defect in the plane of the photonic crystal acts as a waveguide with bounded defect states. Some of these states exist already below the bulk modes and are guided by total internal reflection. They see the photonic crystal as a more or less homogeneous medium with effective refractive index. These modes are very similar to ridge waveguide modes. Some other modes are confined within the line defect due to destructive interferences with the surrounding photonic crystal. This second guiding mechanism is based only on the existence of the photonic band gap and do not exist in a ridge waveguide. If the photonic crystal is made in a thin dielectric slab, light can also be confined in the vertical direction by total internal reflection due to the index contrast between the slab and its surrounding. Combined with the waveguiding along a line defect within the plane of the photonic crystal, a possible 3-D light confinement in 2-D photonic crystals can be achieved.

Acknowledgments

We would like to thank funding by the BMBF within the project HiPhocs. We are also grateful to M. Agio, J. Schilling, F. Müller, T. Geppert, and R. Hillebrand for stimulating discussions.

References

1. John, S., Strong localization of photons in certain disordered dielectric super-lattices, Phys. Rev. Lett. 58, 23 (1987).
2. Yablonovitch, E., Inhibited spontaneous emission in solid-state physics and electronics, Phys. Rev. Lett. 58, 20 (1987).
3. Joannopoulos, J.D., Meade, R.D., and Winn, J.N., *Photonic Crystals, Molding the Flow of Light*, Princeton University Press, Princeton, NJ (1995).
4. Schilling, J., Müller, F., et al., Three-dimensional photonic crystals based on macroporous silicon with modulated pore diameter, Appl. Phys. Lett. 78, 1180 (2001).
5. Birner, A., Wehrspohn, R.B., et al., Silicon-based photonic crystals, Adv. Mater. 13, 377 (2001).
6. Noda, S., Tomoda, K., et al., Full three-dimensional photonic bandgap crystals at near-infrared wavelengths, Science 289, 604 (2000).
7. Blanco, A., Chomski, E., et al., Large-scale synthesis of a silicon photonic crystal with a complete three-dimensional bandgap near 1.5 micrometres, Nature 405, 437 (2000).
8. Benisty, H., Labilloy, D., et al., Radiation losses of waveguide-based two-dimensional photonic crystals: Positive role of the substrate, Appl. Phys. Lett. 76, 532 (2000).
9. Johnson, S.G., Fan, S., et al., Guided-modes in photonic crystal slabs, Phys. Rev. B 60, 5751 (1999).
10. Villeneuve, P.R., Fan, S., et al., Three-dimensional photon confinement in photonic crystals of low-dimensional periodicity, IEE Proc. Optoelectron. 145, 384 (1998).
11. Weisbuch, C., Benisty, H., et al., Advances in photonic crystals, Phys. Status Solidi 221, 93 (2000).
12. Megens, M., Wijnhoven, J., et al., Fluorescence lifetimes and linewidths of dye in photonic crystals, Phys. Rev. A 59, 4727 (1999).

13. Busch, K. and John, S., Photonic band gap formation in certain self-organizing systems, Phys. Rev. E 58, 3896 (1998).

14. Genereux, F., Leonard, S.W., et al., Large birefringence in two-dimensional silicon photonic crystals, Phys. Rev. B 63, 16, 1101 (2001).

15. Kosaka, H., Kawashima, T., et al., Superprism phenomena in photonic crystals, Phys. Rev. B 58, 10, 096 (1998).

16. Park, W. and Summers, C.J., Extraordinary refraction and dispersion in two-dimensional photonic-crystal slabs, Opt. Lett. 27, 1397 (2002).

17. Luo, C., Johnson, S.G., et al., All-angle negative refraction without negative effective index, Phys. Rev. B 65, 201104 (2002).

18. Meade, R.D., Brommer, K.D., et al,. Existence of a photonic band gap in two dimensions, Appl. Phys. Lett. 61, 495 (1992).

19. Johnson, S.G., and Joannopoulos, J.D., Block-iterative frequency-domain methods for Maxwell's equations in a planewave basis, Opt. Express. 8, 173 (2001).

20. Jamois, C., Wehrspohn, R.B., et al., Silicon-based photonic crystal slabs: Two concepts, IEEE J. Quantum Electron. 38, 805 (2002).

21. Lehmann, V., and Föll, H., Formation mechanism and properties of electrochemically etched tranches in n-type silicon, J. Electrochem. Soc. 137, 653 (1990).

22. Lehmann, V., The physics of macropore formation in low-doped n-type silicon, J. Electrochem. Soc. 104, 2836 (1993).

23. Schilling, J., Birner, A., et al., Optical characterisation of 2D macroporous-silicon photonic crystals with bandgaps around 1.5 and 1.3μm, Opt. Mater. 17, 7 (2001).

24. Schilling, J., Wehrspohn, R.B., et al., A model system for two-dimensional and three-dimensional photonic crystals: Macroporous silicon, J. Opt. A: Pure Appl. Opt. 3, 121 (2001).

25. Lončar, M., Doll, T., et al., Design and fabrication of silicon photonic crystal optical waveguides, J. Lightwave Technol. 18, 1402 (2000).

26. Baba, T., Motegi, A., et al., Light propagation characteristics of straight single-line defect waveguides in photonic crystal slabs fabricated into a silicon-on-insulator substrate, IEEE J. Quantum Electron. 38, 743 (2002).

27. Talneau, A., Le Gouezigou, L., and Bouadma, N., Quantitative measurement of low propagation losses at 1.55 μm on planar photonic crystal waveguides, Opt. Lett. 26, 1259 (2001).

28. Weisbuch, C., Benisty, H., et al., 3D control of light in waveguide-based two-dimensional photonic crystals, IEICE Trans. Commun. E84-B, 1286 (2001).

29. Notomi, M., Shinya, A., et al., Structural tuning of guided modes of line-defect waveguides of silicon-on-insulator photonic crystal slabs, IEEE J. Quantum Electron. 38, 736 (2002).

30. Bogaerts, W., Wiaux, V., et al., Fabrication of photonic crystals in silicon-on-insulator using 248-nm deep UV lithography, IEEE J. Selected Topics Quantum Electron. 8, 928 (2002).

31. Qiu, M., Effective index method for heterostructure-slab-waveguide-based two-dimensional photonic crystals, Appl. Phys. Lett. 81, 1163 (2002).

32. Andreani, L.C., and Agio, M., Photonic bands and gap maps in a photonic crystal slab, IEEE J. Quantum Electron. 38, 891 (2002).

33. Qiu, M., Jaskorzynska, B., et al., Time-domain 2D modeling of slab-waveguide based photonic-crystal devices in the presence of out-of-plane radiation losses, Microwave Opt. Technol. Lett. 34, 387 (2002).

34. Ochiai, T. and Sakoda, K., Dispersion relation and optical transmittance of a hexagonal photonic crystal slab, Phys. Rev. B 63, 125, 107 (2001).
35. Johnson, S.G., Villeneuve, P., et al., Linear waveguides in photonic-crystal slabs, Phys. Rev. B 62, 8212 (2000).
36. Leonard, S. W., van Driel, H.M., et al., Single-mode transmission in two-dimensional macroporous silicon photonic crystal waveguides, Opt. Lett. 25, 1550 (2000).
37. Andreani, L.C. and Agio, M., Intrinsic diffraction losses in photonic crystal waveguides with line defects, Appl. Phys. Lett. 82, 2011 (2003).
38. Lončar, M., Nedeljković, D., et al., Experimental and theoretical confirmation of Bloch-mode light propagation in planar photonic crystal waveguides, Appl. Phys. Lett. 80, 1689 (2002).

Interferometric Measurement of Microlenses Including Cylindrical Lenses

Norbert Lindlein, Jürgen Lamprecht, and Johannes Schwider

Lehrstuhl für Optik, Physikalisches Institut der Friedrich-Alexander-Universität
Erlangen-Nürnberg, Staudtstr. 7/B2, D-91058 Erlangen, Germany
nlindlei@optik.uni-erlangen.de

1 Introduction

Passive optical elements are essential parts of microoptical subsystems. The optical function of the elements is brought about either (1) through special surface forms or (2) with corrugated surfaces through the groove profile. Alternatively, also the distribution of the refractive index in volume media in form of graded index distributions (GRIN lenses) or as periodic index variation (holographic volume elements) can be exploited to achieve useful optical light manipulations as, for example, focusing or deflection of light beams.

Consequently, the following physical effects are exploited for achieving the wanted optical performance: (1) refraction, (2) reflection, (3) diffraction, (4) birefringence, and (5) polarization. In the context of microoptical elements, mainly the first three physical effects are the most essential ones.

Therefore, the emphasis will be on the measuring principles for (1) surface forms, (2) profile measurements, and (3) optical performance of the element as a whole.

Since already small surface or refractive-index variations on the order of parts of a wavelength will influence the optical performance, wave-optical methods have the highest priority in this description.

However, some classification might be useful. Therefore the surface shape will be measured in reflected light, avoiding the effect of the volume or of the other surface of a lens on the measuring result. The main interest will, of course, be on the testing of lenses because focusing and collimating are two very essential deflection operations in addition to the light deflection of plane waves. Already the transition of light signals across a certain distance requires a beam expansion or the angular spectrum of a wave emitted by a laser needs some beam shaping in order to keep the losses and cross-talk within the specifications of the experiment. Therefore, it is obvious that lenses play a key role in micro-optical systems.

On the one hand, the optical performance of lenses will be measured in transmitted light since this configuration is very closely related to the final use of the element. On the other hand, the microstructure measurement of diffractive surface elements requires optical profilometers as, for example, white-light interferometers, atomic force microscopes, or Talysurf measurements. In this case, the main emphasis is the correct measurement of the groove profile and not so much the wave aberrations of the element since it will be made by lithographic means where the macrogeometry of the diffractive element will, in general, not be the problem. [1]

Diffractive volume elements in dichromated gelatin or other photorefractive materials can be structured for the purpose of light deflection or focusing by exposing the material to interferometrically produced intensity distributions, where the main topic is the wave aberrations generated by the wavelength mismatch between exposure and final application.

The measurement of such volume elements will require very general interferometric setups and is only of limited importance for microoptical subsystems because of the strong chromatic variations of the performance with the wavelength.

The main emphasis will therefore be on the testing of refractive lenses, which are relatively tolerant to wavelength changes.

Within the framework of testing of "passive microoptical elements," the microoptics concept needs some interpretation since the frontier between microelements and macroelements is more or less unsharp.

In the case of microlenses, it makes sense to discuss this matter from the point of view of the functionality. Typical quantities are diameter and focal length or f-number. The microworld will be governed by the need to connect over rather small distances. If the optical function of a lens is considered, there is an upper limit to such distances given by the Fresnel number, which describes the effect of diffraction. The Fresnel number N is defined as [1]

$$N = \frac{D^2}{\lambda f} , \tag{1}$$

where D is the diameter of the lens, λ is the used wavelength, and f is the focal length or transition distance.

For Fresnel numbers below 1, the diffraction limits the propagation length, which means that typical distances will be on the order of millimeters or centimeters and diameters of the lenses below 1 mm.

On the other hand, the optical power of lenses having only a single curved surface will be limited to a half-sphere plano-convex lens with the focal length

$$f = \frac{D}{2(n-1)} , \tag{2}$$

---

[1] The use of diffractive lenses with very high spatial frequencies might make a difference because the global phase retardation introduced by the diffractive structure is, to some extent, dependent on the frequency itself, which might demand the measurement of the wave aberrations.

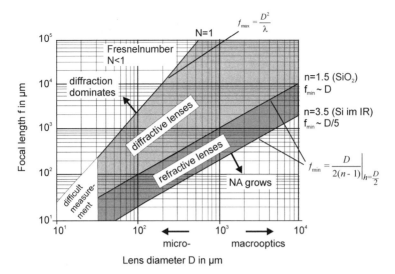

Fig. 1. Feasibility chart for microlenses. Diffractive and refractive microlenses predominantly occupy the regions of low and high numerical apertures, respectively. Quantitative measurements of the wave aberrations become difficult for lens diameters below 20–30 µm. Paraxial focal length estimates are given as limiting cases for quartz glass and silicon.

or the f-number will be limited to

$$f/\# = \frac{f}{D} = \frac{1}{2(n-1)} \; . \tag{3}$$

There will be a transition to the macro region for lens testing for diameters above 1 mm because diffraction effects will become less serious. Diffraction at the rims of the microlens or the small dimensions are one main cause for the necessity to make microlens testing a special subject of research [2].

Testing with the help of wave-optical methods comprises, of course, interferometry for shape and wave aberration measurement, wave-front sensing methods as the Shack–Hartmann or Hartmann sensor, and even confocal microscopy. The selection of the most suitable method will very much depend on the quantity to be measured. We will mainly discern surface deviation measurement, wave aberration measurement, and profilometry.

The necessities for optical testing depend very much on the application field. A coarse classification line can be drawn between the function of light propagation over some distance and the function as a light-collecting element (e.g., in front of an area photodetector or a window of a liquid-crystal display (LCD)).

In the first case, the light has to be collimated or the laser beam has to be shaped in order to bridge some distance or to couple light into a monomode fiber. Such lenses should be diffraction limited; therefore, optical testing on the sub-λ-level will be mandatory, which might mean measurements of the surface shape during the lens forming process and the final control of the wave aberrations after finishing of the lens.

In the second limiting case, where the target area for focusing is very much larger than the Airy disk, it is not in every case necessary to measure with the highest accuracy, which enables also relaxed procedures when measuring wave aberrations or profiles.

The selection of the test method for refractive microlenses depends very strongly on the material of the lens. The main features are transparency to the light and the refractive index. In this connection, let us consider the necessary measuring sensitivity for the surface deviations of spherical surfaces measured with a reflected-light instrument. A surface deviation of Δz causes a wave aberration in transmitted light on the order of

$$\Delta W = (n - 1)\Delta z \ . \tag{4}$$

If the Rayleigh criterion is assumed as the limiting case for diffraction-limited optics, then the allowable surface deviation is on the order of

$$\Delta z \leq \frac{\lambda}{4(n - 1)} \ . \tag{5}$$

Therefore, Si-lenses for the mid-IR will require that the sensitivity of the measuring instrument is much better than $\lambda/10$ (say, e.g., $\lambda/100$), which means that interferometric measurements or comparably sensitive Shack–Hartmann tests are absolutely necessary.

Because of the required rather high sensitivity of the test, mainly null tests will be mandatory, which means the involvement of some null optics. Null lenses within an interferometer have to be free from aberrations down to a level of $\lambda/10$ or even better. The remaining aberrations caused by the null lens and the interferometric components define the accuracy limits of the test setup. Here, we will refrain from the discussion of calibration issues as, for example, absolute sphericity tests [3], since microlenses show deviations from the ideal one order of magnitude above those of a carefully designed interferometer or test setup.

2 Surface Deviation Measurement for Spherical Lenses

The optical function of spherically curved boundaries between dielectrica is also used in the microregion. Several methods have been exploited to produce spherical shapes. To get the production method under control, shape measurements are mandatory. Commonly, tactile profilometers or white-light interferometry are used for this purpose. The problem with this methodology is accuracy. For the optical performance, small deviations on the order of parts

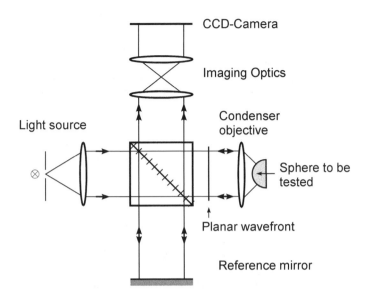

Fig. 2. Twyman–Green interferometer derived from the Michelson interferometer with a specialized test arm containing a beam-shaping condenser objective and the test sample.

of a wavelength can be disastrous. In order to obtain sufficient accuracy, the optical engineer uses null tests; that is, only the small deviations of the surface under test from an ideal surface – here a sphere – are measured. In other words, the surface to be tested is probed by a spherical wave front that matches the surface curvature. To generate such a wave front, high-quality beam-shaping optics is necessary. Fortunately, in the microregion, microobjectives optimized for microscopy fulfill all necessary boundary conditions.

Shape deviation measurement will be made in reflected light in order to separate surface effects from other contributions. The most suitable solution for the test is the Twyman–Green interferometer [3, 4]. A general scheme is given in Figure 2.

The spherical surface is positioned in the test arm and is illuminated by means of a microscopic objective, where the focal point of this objective coincides with the center of curvature of the surface under test. The combination condenser objective surface under test delivers in double passage a nearly plane wave front carrying the surface information as a wave aberration of the wave front. In the detector plane, this wave front is superimposed with a plane reference wave from the reference arm, resulting in an interference pattern representing the surface quality. To ensure highest accuracy, the imaging optics has to image the surface under test sharply.

However, reflected-light interferometers are prone to parasitic fringes from reflections coming from all boundary surfaces in the light path. This is especially true if the highly coherent light from a HeNe laser is used for a light source. The ease in obtaining interference fringes with laser sources has to be paid for with a high coherent noise background limiting the ultimate accuracy of the phase evaluation. However, through the transition to partially coherent illumination, there is, due to fringe localization effects, a severe improvement of the smoothness of the interference fringes possible (see Appendix A).

As a spatially incoherent extended source, a HeNe-laser spot on a rotating scatterer can be used. In this way, a monochromatic incoherent light source can be obtained with the big advantage that the coherence length is very large. The extension of the region along the optical axis with high-contrast fringes can be restricted to dimensions that eliminate parasitic fringes of optical surfaces of all of the auxiliary optics in the interferometer.

Interferometry with partially coherent light requires positioning of the reference mirror into a plane coinciding with the image of the vertex of the lens under testing. This means that the reference mirror has to be shifted along the optical axis into the correct position. In general, this position will not coincide with the position for zero optical path difference, which makes temporal coherence mandatory. The laser spot on a rotating scatterer is therefore a valuable alternative to a spectrum lamp.

High-contrast fringes make a careful intensity balance between the two arms of the Twyman–Green necessary, which is easily attainable by polarizing beam-splitter groups. Such a beam-splitter combination consists of half-wave plates polarizing the beam splitter, quarter-wave plates in the two arms of the interferometer, and a polarizer in the exit of the interferometer. An actual test setup using an extended monochromatic source is shown in Figure 3.

For the evaluation of the phase distribution, phase-shifting interferometry is commonly used (for an explanation, see Appendix B).

The more stringent adjustment requirements compared to coherent illumination with the Twyman–Green interferometer make additional optics for the adjustment of the interferometer necessary. To bring the intermediate image of the lens surface into coincidence with the reference mirror plane, a field stop is used. Auxiliary optics in the entrance of the interferometer images the field stop onto the surface to be tested, which can be controlled in the camera image plane. First, the surface under testing has to be sharply imaged onto the detector. Second the field stop is imaged onto the surface and, third the reference mirror is shifted until the field stop image from the reference arm is also sharp on the detector. Then, the interferometer will show high-contrast fringes. In addition to this precautions, it is also possible to remove the rotating scatterer from the light path to realize coherent illumination. In this case, interference fringe will certainly be observable, which can help to reach the best adjustment fast.

The Twyman–Green test is a null test; that is, it delivers the deviations of the surface from an ideal sphere but does not immediately give information

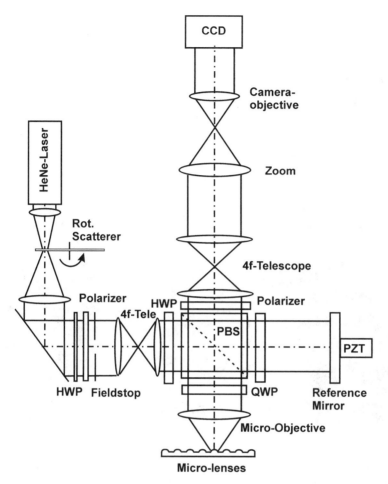

Fig. 3. Twyman–Green interferometer with partially coherent illumination for testing microspheres. The reference mirror has to be shifted along the optical axis to bring the mirror into coincidence with the position of the intermediate image of the surface to be tested formed by the microobjective in the test arm.

on the radius of curvature. This additional information can be obtained by adjusting, in addition, the so-called cat's-eye position by shifting the sphere along the axis until the focus of the condenser coincides with the vertex of the sphere. The distance between the basic and the cat's-eye position (see Fig. 4) is the radius of curvature.

The correct adjustment of the axial positions can be controlled by the planeness of the wave front returning from the test arm. However, due to an inversion of the ray path in the cat's-eye adjustment, the scatterer has to be removed from the illuminating unit in order to obtain interference fringes. An

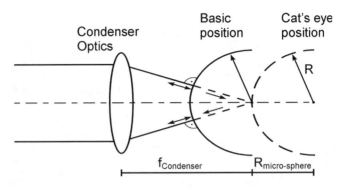

Fig. 4. Radius-of-curvature measurement by shifting the lens from the basic to the cat's-eye position. In both adjustments, straight, parallel, and equidistant fringes will indicate that a plane wave leaves the reference arm.

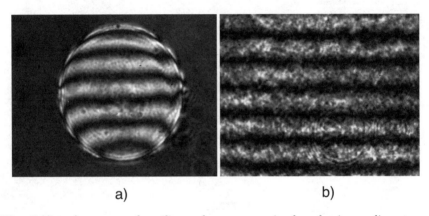

a)   b)

Fig. 5. Interferograms of a silicon plan-convex microlens having a diameter of $250\,\mu m$ and a radius of curvature of $750\,\mu m$. (a) Basic position indicating the deviations from an ideal sphere; (b) interferogram of the cat's-eye adjustment with the rotating scatterer removed.

example for the interference patterns obtained under these circumstances is given in Figure 5.

Figures 6–8 show the results of the test of a lens made with microjet printing. Figure 6 shows the interferogram together with the modulo 2π phase picture following as a first step of the phase-shifting evaluation.

The next step is the unwrapping of the modulo phase pattern. The result of this procedure is given in Figure 7.

It is common usage to fit Zernike polynomials to the measured data in order to obtain the deviations in a smoothed representation. This enables the indication of certain aberrations to the surface defects of the microlens (see Fig. 8).

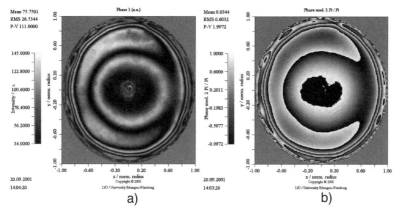

Fig. 6. (a) Interferogram of the surface deviations; (b) mod 2π phase image after phase-shifting evaluation.

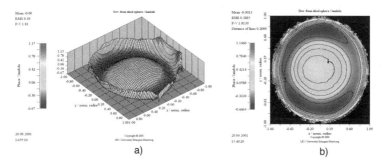

Fig. 7. Deviations from an ideal sphere. (a) Pseudo-3-D plot; (b) contour line plot. Tilt and defocus terms removed from the measured data.

To demonstrate strong and more or less irregular surface defects, a deviation picture of a microlens produced via laser ablation in PMMA is shown in Figure 9.

3 Surface Deviation Measurement of Microcylinder Lenses

Cylinder lenses can be tested in a null test. For this purpose, a diffractive null component can be used for beam shaping or as a reference element. However, microcylinder lenses having a very large numerical aperture (e.g., FAC lenses for beam-shaping applications in front of laser bars) will require rather high spatial frequencies in the diffractive structure (see Appendix C).

The cylinder symmetry suggests a solution resting on the use of two diffractive elements together with a grazing incidence solution onto the surface to be

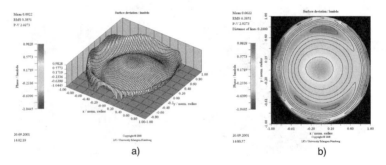

Fig. 8. Deviations in Zernike representation. (a) Pseudo-3 D plot; (b) contour line plot.

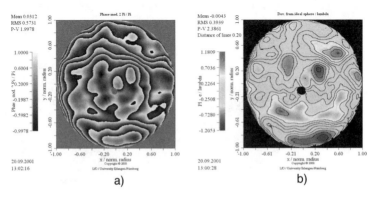

Fig. 9. Deviation picture of a laser-ablated microlens in PMMA: (a) phase mod 2λ; (b) contour line plot, tilt, and defocus terms removed.

tested. Splitting of the functions into a beam splitter and a beam combining diffractive element provides significant advantages: (1) The spatial frequency of the diffractive optical element is constant and can be chosen within the margins of the lithography device, (2) the structure of the element can easily be adapted also to cylinder lenses with an arbitrary mathematically convex meridian curve, and (3) the test interferometer avoids double passages through the diffractive element, avoiding, in this way, parasitic beams. The interferometric setup can be inferred from Figure 10. One diffracted first order is used to illuminate the cylinder lens. The diffractive structure consists of curves parallel to the meridian curve. The zero-order light can be exploited as reference. The second diffractive optical element (DOE) acts as a beam combiner, bringing the waves to interference. Phase shifts for the evaluation of the phase distribution can be introduced by shifting one of the DOE's axially by Δz. The axial shift Δz of one DOE to generate a phase shift by one period of the fringe pattern is

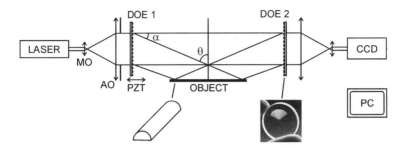

Fig. 10. Scheme of a grazing incidence interferometer for measuring surface deviations of microcylinder lenses. The DOE 1 and DOE 2 are beam splitters and simultaneously beam shaper for the plane wave entering and the conical wave leaving. Lenses with a noncylindrical meridian curve can also be tested by structuring of the DOE accordingly.

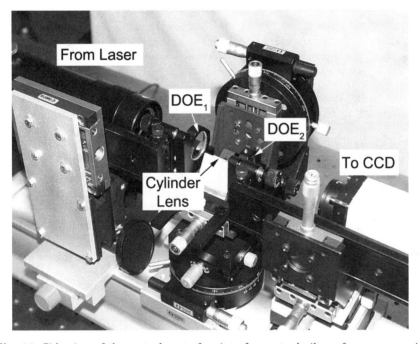

Fig. 11. Side view of the central part of an interferometer built up from commercial adjustment units comprising the two DOEs and the lens under test.

$$\Delta z = 2\frac{p^2}{\lambda} \,, \tag{6}$$

where p is the pitch of the DOE and λ is the wavelength used in the interferometer. An experimental setup based on the scheme of Figure 10 is given in Figure 11.

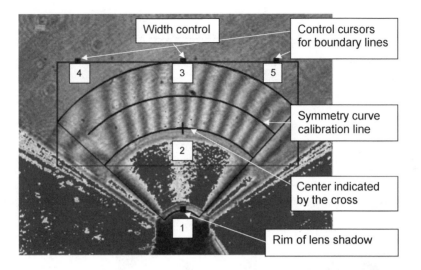

Fig. 12. Interactive mask field for the selection of the region of interest; the direction along the cylinder axis is running upward and the coordinate across runs laterally.

Due to the special geometry, the surface is mapped into a conical-shaped region on the DOE or the charge-coupled device (CCD) chip, respectively (see Fig. 12).

The empty interferometer has to be set up beforehand. For this purpose, the structured DOE regions are larger than the actual need would require. This leads to an overlap of the wave fronts of the zero and the first orders in the rim regions on DOE2 and also produces interference fringes. These fringes can be used to obtain a basic adjustment because the interferometer is correctly adjusted if the fringes in the regions of overlap are "fluffed out." In addition, the DOE designer can place auxiliary structures in the regions outside the test field, which indicate rotations and distances of the two DOEs relative to each other.

The region of interest is selected with the help of the mask field of Figure 12. One has to bring the rim of the lens shadow into coincidence with the lower arc and then adjust the mask field to the interference pattern on the CCD chip.

Figure 13 shows the result of a phase-shift evaluation. The measurement gives the deviations in units of the effective wavelength, where a phase difference of 2π corresponds with a surface deviation of $p/2$, with p being the pitch of the DOE structure (see Appendix C).

With a value of $p = 5\,\mu m$, one can achieve repeatabilities on the order of 25 nm root mean square.

The same result in a contour line representation of the unfiltered data and as a polynomial is shown in Figure 14.

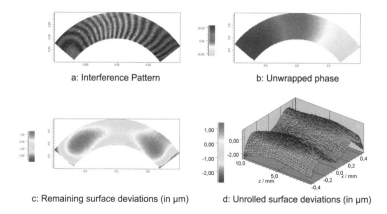

a: Interference Pattern   b: Unwrapped phase

c: Remaining surface deviations (in µm)   d: Unrolled surface deviations (in µm)

Fig. 13. The steps of a surface measurement: (a) Interference pattern with noticeable misalignment in the x direction; (b) unwrapped phase map containing the adjustment errors and the surface information ($P\text{--}V$ value about 40 µm); (c) remaining surface deviations after the reduction step; (d) measurement result in pseudo-3-D view where the mantle surface has been unrolled into the drawing plane. The z axis corresponds the length and the x axis corresponds the width of the lens.

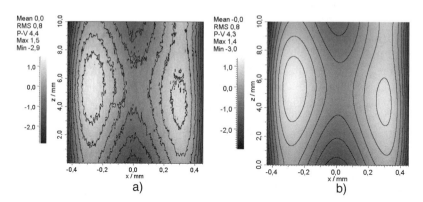

a)   b)

Fig. 14. Unwound surface data in microns as contour line plot (local data (a)) and contour line plot for a polynomial fit of degree 10 (b).

4 Measurement of Wave Aberrations

Lens systems can be tested in reflected-light interferometers (Twyman–Green interferometer) using double-pass as well as in transmitted-light instruments using a single-pass geometry (Mach–Zehnder interferometer).

The first approach has its advantages if high-quality systems are to be tested because the light traverses the optical surfaces of the system under test at the same locations due to the small-angle aberrations of the system under test. A double-pass interferogram is, therefore, a sufficient representation of the optical channel under test.

For microlenses, the situation might be different, because the state of correction will, in general, be poorer because only a single surface can be bent to obtain optical power. Inevitably, the remaining aberrations might be one order of magnitude worse than in the above-discussed case. In addition, with microlenses, the imaging problems of the system under test onto the camera are more involved since, in the general case, the diffracted wave at the rim of the lens will not be negligible. Only if a convex reference mirror can be positioned in the very neighborhood of the lens aperture, these disturbances might be tolerable. The use of a concave reference mirror is not recommended because of the imaging situation. Double-pass setups are therefore prone to systematic errors.

In the case of the test of microlenses, a high-quality microscopic objective of adequate numerical aperture is sufficient since the deviations of the microlenses are greater by at least one order of magnitude.

In addition, transmitted single-pass setups are not so strongly impaired by parasitic fringes because the reflected light originating at glass surfaces in the instrument has to be reflected twice in order to be scattered in the forward direction as the wave used in the test. This reduces the intensity contributions from spurious reflections to the "pars pro mille" region compared with the test waves.

Whereas the Twyman–Green geometry only delivers sufficiently smooth interferograms under partially coherent illumination, the Mach–Zehnder interferometer delivers also under laser illumination sufficiently smooth interferograms, offering in this way greater design freedom since violations of symmetry rules do not impair the contrast of the interference pattern.

Additionally, the single-pass geometry offers for slow lenses even the opportunity to measure the total phase lag of the microlenses without violations of the sampling theorem. This makes an estimation of the back focal length possible.

The wave aberrations are the optical excess paths from the path an ideal spherical wave would take (see Fig. 15) from the exit pupil to the image point.

Here, we will restrict the discussion to the configuration ∞/f (i.e., plane-wave illumination of the microlens and, consequently, with the image point in the back focal plane). After the wave aberrations have been determined, the point spread function and the optical transfer function can be derived through Fourier transformation for the selected field point. In most cases, the axial focus point will be selected for symmetry reasons.

The wave aberrations will be measured for refractive lenses in transmitted light. Therefore, a Mach–Zehnder geometry is used where a telescopic microscope is incorporated in the object light path and the reference wave front is a plane wave traveling along the reference path of the Mach–Zehnder.

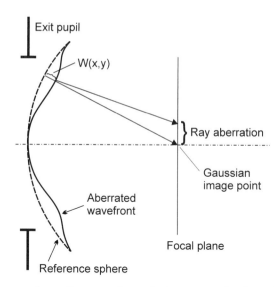

Fig. 15. The wave aberrations are the optical excess paths from the path an ideal spherical wave would take.

The plane-wave philosophy reduces the systematic error contributions in the interferometer to the optical tolerances of the interferometer components.

One of the boundary conditions has been that rather large lens arrays should be measurable which requires a large x, y-stage and, in addition, a perpendicular support for the lens arrays. This has been the reason for implementing the Mach–Zehnder geometry [5, 6] into a commercial microscope [Carl Zeiss Jena (JENATECH)]. This could only be realized using a fiber-based solution (see Fig. 16), which will now be described.

The beam of a HeNe laser is split by a polarizing beam splitter group consisting of a polarizing beam-splitter cube (PBS) and a half-wave plate (HWP) at the entrance to enable the balancing of the two interfering beams. The fibers are of the polarization-maintaining type in order to keep the polarization signature for the balancing of the brightness in the two beams of the interferometer.

The second beam splitter is a nonpolarizing one. In order to force the two beams to interfere, the plane of polarization of the reference beam is rotated by 90° just by twisting the fiber end accordingly, which can be controlled with the help of a polarizer. This solution has been chosen in order to avoid the use of a sheet polarizer in front of the CCD chip. It turned out that the optical quality of these polarizers is inferior to the chosen solution. The disadvantage of this solution is a loss in the freedom to continuously adapt the intensity to the dynamic range of the CCD array. To obtain the necessary degree of freedom for the intensity control, two polarizers are placed in front of the

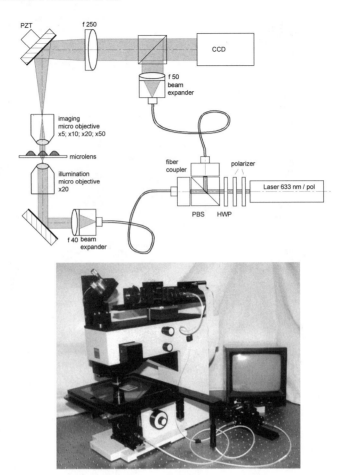

Fig. 16. Scheme of the Mach–Zehnder interference microscope for measuring the wave aberrations of microlenses. The Mach–Zehnder interferometer is based on a Zeiss-microscope type JENATECH.

HWP. The polarizer in front of the HWP stays fixed and the second polarizer can be used as an attenuator for the laser intensity.

The microlens to be measured is illuminated by means of a spherical wave from the back side, which is generated by means of a high-quality microobjective. The microlens transforms this wave back into a plane wave. A telescopic microscope produces an image of the lens on a CCD camera, on the one hand, and an expanded plane wave on the other, which enters a second polarizing beam splitter is superimposed by a plane reference wave fed through the second fiber and an expansion lens to a second beam splitter. One of the mirrors in the object arm of the Mach–Zehnder interferometer (MZ) is mounted on a piezo transducer (PZT) driver to enable phase-shifting interferometry. For

Fig. 17. Interferogram of a diffractive eight-level microlens.

this purpose, the CCD camera has a pixel-synchronous frame grabber coupled to a PC.

The interference microscope is equipped with two "probe wave-front" alternatives: one for weak-phase objects and the other for wave-aberration measurements.

5 Weak-Phase Objects

For general phase objects, plane-wave illumination is most suitable. Typical objects are biological objects and, to some extent, also binary objects as diffractive optical elements as long as the phase lag between neighboring pixels stays well below π. Such an example is given in Figure 17, where one period of an eight-step Fresnel lens is shown as an interferogram.

Another example is a refractive beam-shaping element which is several wavelengths deep. It is known that beam shaping can be achieved through diffractive optical elements. However, in the case of an unequivocal spatial relationship between the planes of the element and the reconstruction plane, one may even use refractive beam-shaping elements, as is done in the case of microlenses or homogenizers. Since one can assume that the elements are weak-phase elements, plane-wave illumination is suitable to determine the total phase lag due to the profile of the refractive beam shaper. A typical sample has been measured with the help of the Mach–Zehnder microscope. The results are given in Figure 18.

Even for slow microlenses, the total phase lag introduced by the lenses can be measured if the illuminating wave is a plane wave. In this case, one can obtain the test result for a subsection taken from the array of microlenses (see Fig. 19). This might be of some value for uniformity tests of microlens arrays.

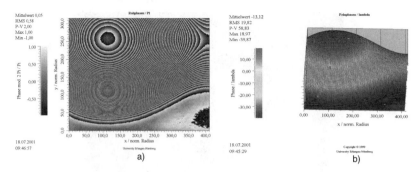

Fig. 18. Phase distribution of a refractive beam shaper measured in transmitted light (selected field from a symmetrical structure only): (a) phase mod 2π; (b) pseudo-3-D plot of the phase distribution (scale in wavelengths at 633 nm).

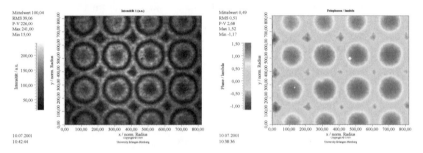

Fig. 19. Array of microlenses illuminated with a plane wave.

In addition, it is possible to estimate the power of the lens if the diameter of the lens is known and the number of fringes is counted.

If for this estimation the assumptions of a paraxial imaging are presumed, the power

$$D = \frac{1}{f} \tag{7}$$

can be calculated from the radius ϱ of the lens aperture and the order number of the fringe pattern in the rim region (i.e., the number of fringes m counted from the center to the rim of the lens):

$$\frac{1}{f} = \frac{2m\lambda}{\varrho^2}. \tag{8}$$

If a phase-shift evaluation is possible, the phase at the rim can be taken, or for fast microlenses, the central part of resolved fringes can be evaluated. Before the phase-shifting evaluation starts, the ring mask being adjusted to the lens aperture can be used because the number of pixels for the diameter is indicated. A simple calibration with the help of a commercial object micrometer delivers the absolute diameter $2\varrho_{\mathrm{mask}}$ in micrometers. With the measured phase Φ in radians one obtains

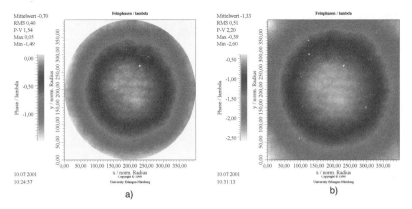

Fig. 20. Total phase lag of a diffused microlens with a quadratic format: (a) circular field; (b) total aperture.

$$\frac{1}{f} = \frac{\Phi\lambda}{\pi \varrho_{\text{mask}}^2}. \tag{9}$$

The removal of linear and quadratic phase terms delivers the wave aberrations also for weak lenses (see Fig. 20). The phase-shifting software delivers such data fields automatically in the form of deviations from the reference sphere or as Zernike coefficients.

6 Microlenses as Strong-Phase Objects

All refractive or diffractive elements which form a spherical wave from a plane wave and vice versa can be tested with the help of the spherical wave-front illumination in the object beam. The element to be tested transforms the illuminating spherical wave into a plane wave, which can be observed by the detector array (lens examples are given in Fig. 21).

In other words, what are measured with the help of the interferometer are the wave aberrations of refractive or diffractive microlenses. From these data, one can calculate the point spread function of the microlens via a fast Fourier transformation (FFT). An additional inverse FFT of the point spread function provides the optical transfer function (OTF), the modulus of which is the modulation transfer function (MTF), which is another essential merit function of optical channels.

The measurement of the wave aberrations is carried out by using the phase-shifting algorithm described in Ref. [7]. The result of the evaluation is the phase distribution in the exit pupil:

$$\Phi(x, y) = \frac{2\pi}{\lambda} W(x, y), \tag{10}$$

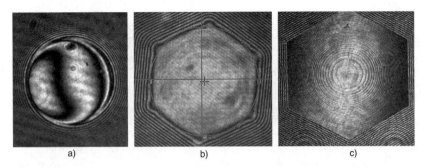

Fig. 21. Interferograms showing the wave aberrations for refractive microlenses: (a) circular aperture and (b) hexagonal aperture; (c) a diffractive microlens.

where $W(x,y)$ are the wave aberrations of the lens under test. In Figures 22 and 23, measuring results are given for quartz glass lenses produced by different techniques.

These wave aberrations can be expanded into Zernike polynomials $W_{\text{Zernike}}(\varrho, \varphi)$ of the form

$$W_{\text{Zernike}}(\varrho, \varphi) = \sum_{n=0}^{G} \sum_{l=0}^{|l| \leq n; (n-|l|) \text{ even}} Z_{nl} Z_n^l(\varrho, \varphi)(\varrho, \varphi), \tag{11}$$

where Z_{nl} are Zernike coefficients,

$$Z_n^l(\varrho \cos\varphi, \varrho \sin\varphi) = R_n^l(r)\, e^{il\varphi},$$

$$R_n^{\pm m}(r) = \sum_{s=0}^{\frac{(n-m)}{2}} (-1)s\, \frac{(n-s)!}{s!\left(\frac{(n+m)}{2}-s\right)!\left(\frac{(n-m)}{2}-s\right)!}\, r^{n-2s},$$

for $4 \leq G = 10$, $|l| \leq n$, $(n-|l|)$ even $m = |l|$.

The point spread function follows then from the Fourier transform of

$$G(x,y) = \exp(i\Phi(x,y)), \tag{12}$$

where the point-spread function (PSF) is

$$\text{PSF}(\xi, \eta) = |\text{FT}\{G(x,y)\}|^2 \tag{13}$$

The optical transfer function is calculated from the PSF data via an inverse Fourier transformation:

$$\text{OTF}(\nu, \mu) = \text{FT}^{-1}\{\text{PSF}(\xi, \eta)\}. \tag{14}$$

In general, the most essential quantity is the MTF, being the modulus of the OTF. The phase of the OTF is in most cases not so conclusive concerning the optical behavior. This is especially true if the wave aberrations

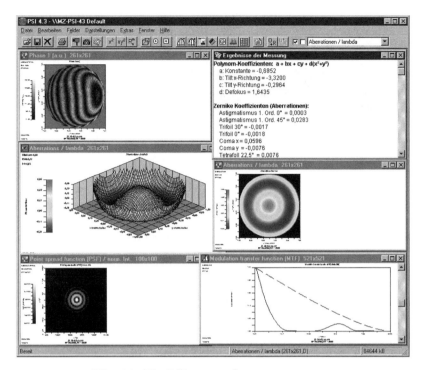

Fig. 22. The PSI screen after a measurement.

are rotational symmetric, which is the most common situation because of the production methods of microlenses. That is the reason why we follow here the general procedure of lens design programs and put the emphasis on the PSF and the MTF.

It can be shown that the OTF is the autocorrelation function of the pupil function [8], which has severe consequences for the domain in the pupil space, which has to be preserved for the calculation of the OTF; that is, the data space should be greater than twice the area of the exit pupil. Calculating the Fourier transform on a PC means the transition to discrete sets of data, which is necessary because the measured data are also given as a discrete matrix of Φ values. Fourier transformations are using the Cooley–Tuckey algorithm (FFT) which relies on datasets in the x and y directions with dimensions being powers of 2 [9].

The usual procedure is as follows: One starts with the matrix of measured data having the dimension (n, n) and fills the dataset with zeros up to the next close power to the basis of 2, and, afterward, this area is again doubled and filled with zeros in such a way that the pupil function fills the central part of the data window. Doing so, it is guaranteed that the autocorrelation function can be calculated. In addition, widening the pupil aperture by adding zeros in the rim region of the pupil provides a much denser sampling in Fourier space,

200   Norbert Lindlein et al.

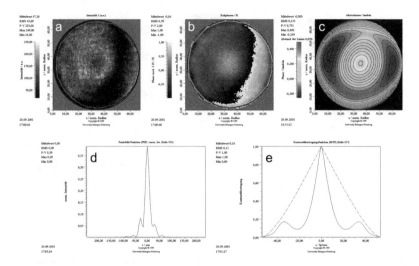

Fig. 23. Interferogram and evaluation results for a spherical microlens made with microjet technology in PMMA: (a) Intensity of the interferogram, (b) wrapped phase map, (c) deviation contour line plot as Zernike fit of degree 10, (d) PSF plot, and (e) MTF plot.

which means, in this case, in the PSF space. This is essential because the PSF only differs from zero within a very small region of the data field because it is the δ response of the optical system under test.

From the Fourier transform and the inverse transform, only relative values follow. In order to obtain absolute figures for the extension of the PSF and the MTF metric quantities have to be entered into the program. These values are the focal length f and the diameter 2ϱ of the lens. In this approximation, it is assumed that the numerical aperture is

$$\mathrm{NA} = \frac{\varrho}{f} \, . \tag{15}$$

It should be mentioned that this expression is only valid if the sine condition is fulfilled or if the principal "plane" H' is a sphere centered about the image point (here, the focus due to the assumed test geometry). Thus, it is advisable to be careful in the case of high-numerical-aperture lenses, which can only be realized with materials having a very high refractive index. This is out of the question in the case of normal fused silica lenses.

For the diameter, one has to take the actual lens aperture, not the filled-in aperture due to the FFT algorithm. This is also true for the MTF since the field used in the transformations is larger than unconditionally necessary. The theoretical cutoff frequency for the ideal lens is

$$\nu_{\mathrm{cutoff}} = \frac{2\mathrm{NA}}{\lambda}. \tag{16}$$

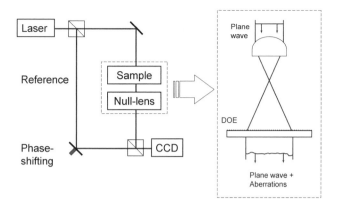

Fig. 24. Test setup for cylindrical lenses in transmitted light with the use of a diffractive optical element, as reference element, and beam shaper (on the left). Actual compensating scheme of the cylindrical wave front with the help of a diffractive reference element which produces a plane wave from the cylindrical wave originating from the FAC lens.

The algorithm yields numerical values normalized by this cutoff frequency.

7 Test of Cylindrical Microlenses in Transmitted Light

Cylindrical microlenses with a high numerical aperture (so-called FAC lenses) are used as beam-shaping elements for the fast axis of laser bars. Numerical apertures as high as 0.8 are quite common. For small M^2 values diffraction-limited performance is required. Therefore, the measurement of the wave aberrations of such lenses is mandatory. Diffraction-limited performance can, in general, only be achieved by aspherization; that is, the surface profile deviates in this case very significantly from an ideal circle.

Wave aberrations can only be measured if the cylindrical curvature of the wave field generated by the cylindrical lens is compensated for by a diffractive optical element. Here, the need for the application of DOEs becomes obvious (see Fig. 24).

One of the problems with the use of DOEs as reference elements for optical testing is their strong chromatic aberration if the wavelength of the test setup differs from the design wavelength. That is the reason why many test setups including DOE references rely on the HeNe laser because it provides both wavelength stability and sufficient coherence length. However, the pump wavelength for solid-state lasers generated by semiconductor laser bars are located in the near-infrared (NIR) region at 808 or 940 nm. Testing the refractive cylinder lenses at these wavelengths or at least in the neighborhood of the pump wavelengths will be necessary (an exemplary interferometer setup is

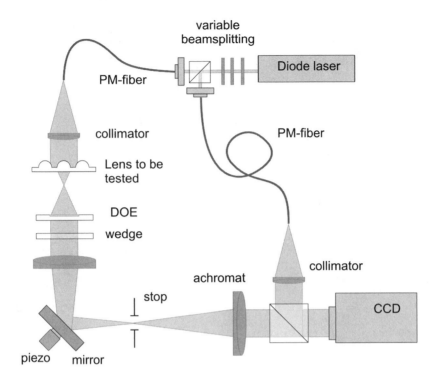

Fig. 25. Actual Mach–Zehnder test interferometer using monomode fiber coupling to the two interferometer arms, enabling a very flexible design.

given in Fig. 25). In addition, the transition to wavelengths in the NIR helps the use of DOEs because the diffractive structures scale with the wavelength.

However, the requirement for the coincidence of design and test wavelength remains. The tolerances for wavelength deviations are very small, which can be shown by calculating the chromatic effects. Figure 26 shows the wavefront aberration which is introduced by a mismatch of only 1 nm between the design and test. Thus, it becomes obvious that a stabilized semiconductor laser with a sufficiently known wavelength is mandatory to get reliable wave-aberration data.

For test applications, the DOE structure can be binary because it is possible to spatially filter the wave field after passing the combination lens/DOE beam shaper. The basic idea is to use plane waves whenever possible, which means that a plane wave should enter and leave the lens/DOE combination apart from a small deviation due to the small deviation from the ideal lens shape.

To enable spatial filtering, a linear phase function along the cylinder axis is added to the phase function of the cylinder DOE. This introduces a tilt to

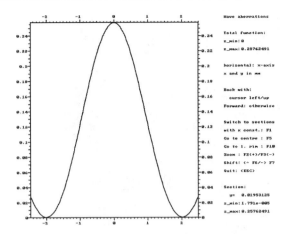

Fig. 26. Wave aberration of the order $\lambda/4$ generated through a wavelength mismatch of only 1 nm from 808 to 807 nm.

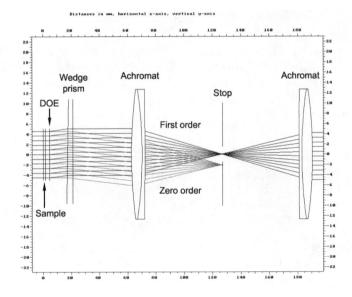

Fig. 27. Simulation of the test arm configuration using a carrier frequency DOE.

the diffracted wave. In order to force the test wave front on-axis, a small prism is added to the ray path (see Fig. 27). In this way, the test wave is on-axis and all disturbing waves can be stopped out being off-axis.

However, in every test setup, be it an interferometer or a Shack–Hartmann sensor, there is the problem of alignment aberrations due to misalignments of the sample under test relative to the frame of the test setup, which also comprises the DOE reference. This is well known from Twyman–Green inter-

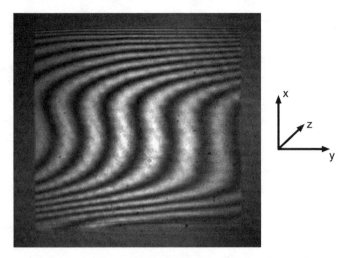

Fig. 28. Interferogram of an acylinder lens showing the effects of astigmatic aberrations due to a rotation about the y axis.

ferometers and the Mach–Zehnder for spherical lens testing. In general terms, the wave aberrations ΔW due to misalignments represented by a shift vector $\varepsilon = (\delta x, \delta y, \delta z)$ are in a linear approximation proportional to the gradient of the wave front $W(x, y)$ of the lens under test:

$$\Delta W(x, y) = \text{grad } W \varepsilon \, . \tag{17}$$

Here, the rotational symmetry is broken, therefore, the paraxial functional representing the misalignment of the cylinder lens for small alignment errors and small numerical apertures is an (x, y)-polynomial of the following form:

$$\Delta W(x, y) = a_1 + a_2 x + a_3 y + a_4 xy + a_5 x^2 + a_6 y x^2 + a_7 x^3, \tag{18}$$

where the cylinder axis is assumed as being parallel to the y axis. In addition, higher orders have to be included in the polynomial for lenses with high numerical apertures:

$$\Delta W(x, y) = a_1 + a_2 x + a_3 y + \frac{(a_4 + a_5 x + a_6 y + a_7 xy)}{\sqrt{f^2 + x^2}}, \tag{19}$$

where f is the focal length of the DOE.

The acylindrical lens has 5 degrees of freedom of movement relative to the frame of the test setup. Astigmatic terms will, for example, occur if the axes of lens and DOE are skew to each other (see Fig. 28). Quadratic contributions will arise for the x coordinate from defocus alignments.

One special problem of microcylinder lens testing are the lateral extensions of the lens on the order of $1\,\text{mm} \times 10\,\text{mm}$ because the slitlike aperture does

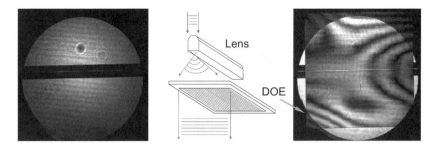

Fig. 29. Combination of a cylinder lens with a DOE in such a way that an anamorphic compensation occurs, which supports the interferometric evaluation of the wave aberrations of the lens. On the left: cylindrical lens without DOE, on the right: anamorphotic rectification of the lens aperture due to the DOE.

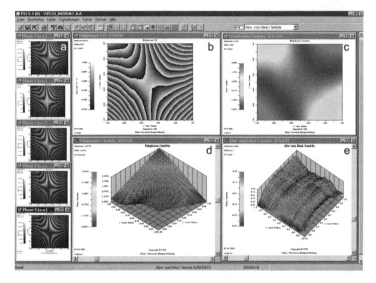

Fig. 30. Test result obtained with the DOE-assisted Mach–Zehnder interferometer of Figure 24. The lens aberrations (e) are extracted from the interferogram by removing alignment contributions. The figure shows the single steps from gathering five phase-shifted intensity pictures (a), the calculation of the phase $\mod 2\pi$ (b), the phase unwrapping (c), and the elimination of alignment aberrations (d).

not match the geometry of the CCD camera. The number of measured values is therefore very much reduced, which hampers the resolution for sections perpendicular to the cylinder axis. An anamorphic system such as a prism telescope could be used to stretch or quench one dimension in order to rectify the geometric mismatch. However, this would mean additional elements causing systematic wave aberrations. A more straightforward solution is the exploitation of the DOE also as an anamorphot as shown in Fig. 29.

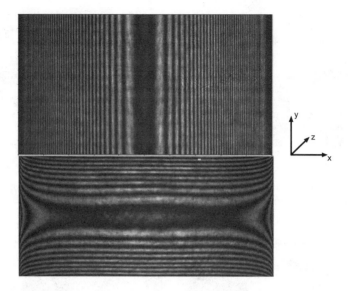

Fig. 31. Top: Illumination with a plane wave; bottom: Illumination with a high-aperture microobjective; cylinder axis is oriented along the y axis.

The lens to be tested could be imaged through the DOE, but there will be astigmatism in the imaging process, introducing a certain uncertainty in the location of the deviations on the lens surface. If the DOE surface is imaged sharply, this astigmatism is avoided, but a good alignment is then mandatory in order to avoid problems with the interpretation of the wave-aberration pictures.

Figure 30 shows a measuring result of a FAC lens with a numerical aperture of 0.8. The single steps from gathering the phase-shifted intensity pictures to extracting the relevant surface information can be inferred from the figure.

Since it is rather difficult to produce DOEs having high spatial frequencies with sufficient accuracy and efficiency, it is useful to think of testing alternatives. The cylinder symmetry of the sample under test enables at least the test along a single section through the profile if the illumination with the help of a high-numerical-aperture microscope objective is chosen. Then, one has, in addition to the central section along the cylinder axis, also information on the deviations from the true cylinder wave in one section of the lens. Global deviations of the vertex region of the cylinder lens can be derived with the help of plane-wave illumination of the lens (see Fig. 31).

The successive measurement of sections perpendicular to the cylinder axis could also give a sufficient survey of the lens performance together with the evaluation under plane-wave illumination for the cylinder vertex region.

Appendix A: The Reduction of Coherent Noise in Interferograms

Interference patterns generated with lasers are plagued by "dust diffraction patterns" and spurious fringes due to the high spatial and temporal coherence. The temporal coherence is, in most cases, advantageous because it alleviates the optical layout of the interferometer and eases the adjustment of the whole setup. However, the coherent noise caused by the high spatial coherence limits the ultimate phase definition to values around $\lambda/50$.

Dust diffraction patterns are enhanced due to the heterodyne advantage of interferometry if spatially coherent illumination with a laser is used.

In the past, before the advent of the laser, interferometry had to use incoherent spectrum lamps. Due to an extended monochromatic light source, the fringe contrast in plane mirror interferometers is localized in the neighborhood of a plane perpendicular to the mean light direction containing the axis of rotation for the light source space; that is, in case of a Michelson interferometer, the crossover line of the two end mirrors of the interferometer [10,11]. In this case, the fringe pattern shows smooth fringes, indicating a disturbance-free phase distribution. Spurious fringes will vanish because their contrast maximum is positioned in another depth region of the interferometer space. Since there is a localization plane, the surface to be tested has to coincide with this plane and should be sharply imaged onto the detector. It happens that this coincidence is realized if the extended light source is at infinity or, equivalently, if the light beam is collimated.

A simple explanation for the localization of the high-contrast region to the condition OPD $= 0$ follows from the path difference of a plane-parallel mirror interferometer:

$$\text{OPD} = 2d\sqrt{n^2 - \sin^2\alpha}, \tag{20}$$

where d is the thickness of the plane-parallel plate, n is the refractive index, and α is the angle of incidence of a light wave. The variation of the OPD with the aperture angle will therefore tend to zero if $d = 0$, which is just the above-mentioned condition.

Hansen showed, in a series of articles, how this principle can be applied to the Twyman–Green interferometer [12–14]. In this case, only the combination of the condenser system plus the spherical surface under test could be replaced by a plane mirror at the position of the intermediate image of the surface under test produced by the condenser system (see Fig. 32).

From this follows that the reference mirror should be axially shiftable in the reference arm and should coincide with the virtual location of the intermediate image (see Fig. 33).

Partially coherent illumination using a laser can be realized by using an extended laser spot on a rotating scatterer. Within the integrating time of the detector, the speckle patterns generated by the scatterer are moving, resulting in an integration effect.

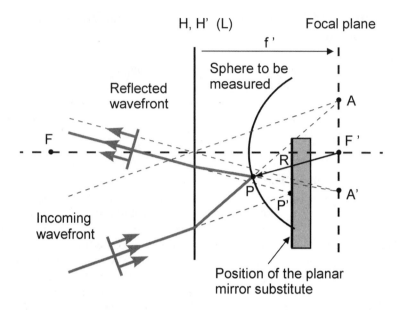

Fig. 32. Position of the plane mirror substitute for the combination condenser system plus spherical surface under test.

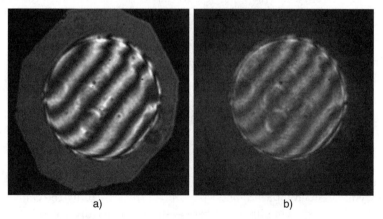

a)   b)

Fig. 33. Interference pattern of the lens under test in the basic position with two different positions of the reference mirror: (a) reference mirror coincides with the intermediate image (field stop image in focus!); (b) reference mirror 50 mm outside the correct position (field stop image unsharp!).

This method opens also the opportunity to remove the scatterer out of the ray path, providing also the case of coherent illumination. The latter possibility is essential if the radius of curvature is to be measured [15] since in the

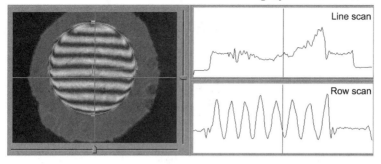

Fig. 34. Interactive "live-video display" of the interference fringe field on the CCD camera.

cat's-eye position, an inversion at the optical axis occurs, which presupposes total spatial coherence.

Appendix B: Evaluation of Phase Distributions Using the Phase-Shifting Method

The measurement of wave-front aberrations using the phase-shift technique [7] requires the intake of several intensity pictures of the interference pattern (s. Fig. 34) differing by the arbitrary reference phase. A minimum of three phase values within one period of the interference pattern is necessary in order to separate the phase from the mean intensity and the visibility. An algorithm immune against calibration errors uses five phase steps [16] with a step width of $\pi/2$ resulting in the raw phase:

$$\Phi = \arctan \frac{2(I_2 - I_4)}{I_1 - 2I_3 + I_5} \mod 2\pi,\tag{21}$$

where the intensities I_1 through I_5 are read in.

These phase values show phase jumps by 2π which have to be removed through next-neighbor operations. The assumption is, of course, that the sampling theorem is fulfilled, which means that in the case of two-beam interference, more than two samples per period of the interference pattern are taken everywhere in the exit pupil of the interferometer.

After the unwrapping operation, the phase values have to be freed from contributions due to the adjustment. In the case of a spherical surface, this is tilt and defocus, which is represented for small misalignments by the following second-order polynomial:

$$\Phi_{\text{def}} = a + bi + cj + d(i^2 + j^2),\tag{22}$$

where i and j are the pixel numbers in the line and row of the CCD array and $a, b, c,$ and d are parameters describing the misalignment. The determination of the latter parameters follows from a least-squares optimization of

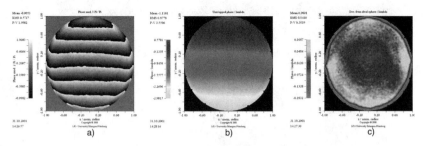

Fig. 35. (a) Phase mod 2π, (b) unwrapped phase distribution, (c) surface deviations from a best-fitting sphere.

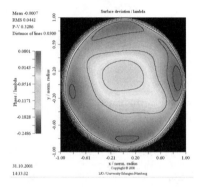

Fig. 36. Contour line plot of the deviation data (Zernike polynomial of degree 10).

the variance:

$$\mathrm{Var} = \sum_{i,j} \left\{ \Phi_{\mathrm{meas}}^{i,j} - \Phi_{\mathrm{def}}^{i,j} \right\}^2 \Rightarrow \mathrm{Min}. \tag{23}$$

With the help of this set of parameters, obtained the measured values are corrected (see Fig. 35c).

The global deviations of the surface can be extracted through fitting of Zernike polynomials to the measured data and the graphic display of this polynomial in form of a contour line plot (see Fig. 36).

Appendix C: Grazing Incidence Interferometry Using Diffractive Optical Elements

The test of aspherics in a null test configuration requires either special null lenses made from refractive components or the use of computer-generated holograms or of DOEs. In the latter case, the wave is generated through diffraction [17]. Common aspherics are tested in perpendicular incidence; that is, the wave normal of the probing wave front impinges along the surface normal of the surface under test (see Fig. 37a).

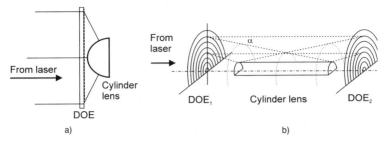

Fig. 37. (a) test with perpendicular incidence of the probing wavefront requires very high spatial frequencies to realize deflection angles, (b) by splitting the function the DOE structure has a constant pitch and the structures are parallel curves to the meridian

In the case of the grazing incidence interferometer using two DOEs, the zero diffraction order is exploited as the reference wave, and one of the diffracted first order waves is exploited as a probe wave front for the surface under test [18,19] (see Fig. 37b). After reflection at the surface, the wave is retransformed into a nearly plane wave front carrying the surface information as wave-front deviation from a plane wave and giving rise to the test interferogram [19–24].

Furthermore, the interferometer is achromatic due to the diffractive splitting units. This means that the chosen wavelength influences only the degree of the anamorphic distortion, not the sensitivity to surface deviations. The interferometer works with an effective wavelength equal to the pitch of the DOEs. This follows from the fact that the diffraction angle α follows from the grating equation

$$\sin \alpha = \frac{\lambda}{p} \tag{24}$$

and the fact that the cosine of the incidence angle is equal to $\sin \alpha$.

Therefore, the simple equation holds:

$$\lambda_{\text{eff}} = p \, . \tag{25}$$

References

1. Saleh B. and Teich, M., *Fundamentals of Photonics*, Wiley–Interscience., New York (1991).
2. Lohmann, A., Justification of Young's diffraction theory, in *Optical Information Processing*, 3rd ed., Lec. Uttenreuth, Germany (1986) p.91.
3. Schulz, G. and Schwider, J., Interferometric testing of smooth surfaces, in *Progress in Optics XIII*, edited by Wolf, E., Elsivier New York (1976).
4. Twyman, F., *Prism and Lens Making*, 2nd edition, Hilger, London (1988).

5. Sickinger, H., Schwider, J., and Manzke, J., Fiber-based Mach–Zehnder interferometer for measuring wave aberrations of microlenses, Optik 110, 239–243 (1999).

6. Sickinger, H., Charakterisierung von sphärischen und zylindrischen Mikrolinsen mit einem Mach–Zehnder-Interferometer, PhD thesis, University, Erlangen (2001).

7. Schwider, J., Advanced evaluation techniques in interferometry, in *Progress in Optics XXVIII*, Elsevier New York (1990), pp. 271–359.

8. Born, M. and Wolf, E., *Principle of Optics*, 7th (expanded) ed., Cambridge University Press, New York (1999).

9. Brigham, O., *FFT – Fast Fourier Transform*, Prentice-Hall, Englewood Cliffs, NJ (1974).

10. Schulz, G., Zweistrahlinterferenz in Planspiegelanordnungen I und II, Opt. Acta 11, 43, 131–143 (1964).

11. Schulz, G. and Schwider, J., Zweistrahlinterferometer, Lichtquellenbildtransformation als Schraubung und neue Effekte an Interferenzstreifen, Optik 21, 587–597 (1964).

12. Hansen, G., Die Sichtbarkeit der Interferenzen beim Michelson- und Twyman-Interferometer, Zeiss-Nachrichten, 4, 109 (1942).

13. Hansen, G., Die Sichtbarkeit der Interferenzen beim Twyman-Interferometer, Optik 12, 5 (1955).

14. Hansen, G. and Kinder, W., Abhängigkeit des Kontrastes der Fizeau-Streifen im Michelson-Interferometer vom Durchmesser der Aperturblende, Optik 15, 560 (1958).

15. Schwider, J. and Falkenstörfer, O., Twyman–Green interferometer for testing microspheres, Opt. Eng. 34, 2972–2975 (1995).

16. Schwider, J., Burow, R., Elssner, K.-E., Grzanna, J., Spolaczyk, R., and Merkel, K., Digital wave-front measuring interferometry: some systematic error sources, Appl. Opt. 22, 3421–3432 (1983).

17. Schwider, J., Interferometric tests for aspherics, in Fabrication and Testing of Aspheres, Vol. 24 of OSA Trends in Opt. Photonics Series, Lindquist, A., Piscotty, M., and Taylor, J., (eds.), Optical Soc. of America (1999).

18. Schwider, J., Verfahren und Anordnung zur Prüfung beliebiger Mantelflächen rotationssymmetrischer Festkörper mittels synthetischer Hologramme, DDR Patent WP 106 769 (filed 4 Jan. 1972).

19. Dresel, T., Schwider J., Wehrhahn, A., and Babin, S., Grazing incidence interferometry applied to the measurement of cylindrical surfaces, Opt. Eng., 34, 3531–3535 (1995).

20. Brinkmann, S., Dresel, T., Schreiner, R., and Schwider, J., Axicon-type test interferometer for cylindrical surfaces, Optik 102, 106–110 (1996).

21. Lindlein, N., Schreiner, R., Brinkmann, S., Dresel, T., and Schwider, J., Axicon-type test interferometer for cylindrical surfaces: systematic error assessment, Appl. Opt. 36, 2791–2795 (1997).

22. Brinkmann, S. Schreiner, R., Dresel, T., and Schwider, J., Interferometric testing of plane and cylindrical workpieces with computer generated holograms, Opt. Eng. 37, 2508–2511 (1998).

23. Dresel, T., Brinkmann, S., Schreiner, R., and Schwider, J., Testing of rod objects by grazing incidence interferometry: Theory, J. Opt. Soc. Am A 15, 2921–2928 (1998).

24. Brinkmann, S., Dresel, T., Schreiner, R., and Schwider, J., Testing of rod objects by grazing incidence interferometry: Experiment, Appl. Opt. 38, 121–125 (1999).

The Integrated Microlens: The Development of Vertical and Horizontal Integration

Masahiro Oikawa

Nippon Sheet Glass Co., Ltd Information & Telecommunication device Division
5-8-1 Nishi-hashimoto Sagamihara, Kanagawa 229-1189 Japan
MasahiroOikawa@mail.nsg.co.jp

1 Introduction

Recent progress of optical fiber communication systems bring about tremendous possibilities of digital communication. In these optical fiber communications systems, lenses play an important role for coupling the light between optical devices such as LDs (laser diodes), fibers and PDs (photodetectors). In order to expand the possibilities of light coupling, lenses are integrated with multiple functions. In this chapter, I introduce the integration technologies of microlenses. The integration technologies can be categorized into two classes: the vertical integration and the horizontal integration. The vertical integration means to integrate an additional function on the lens along its optical axis. The horizontal integration means to integrate microlenses as a planar arrayed structure.

For the vertical integration, dielectric filter coating for the high-performance optical filter is introduced. Also, the planar microlens array that is the horizontally integrated two-dimensional microlens array is introduced. The planar microlens is used in the optical interconnection of an LD array and an optical fiber array. These microlens integration technologies are bringing new possibilities in the optical fiber communication systems.

2 Selfoc (GRIN) Lens

Usually, lenses that focus light uses curved surfaces where the refractive index is changed discontinuously. On the other hand, we sometimes see a mirage in the nature, which has interesting optical phenomena. We see the mirage when the refractive index of the air is gradually changing, which is caused by the temperature change of the air.

In 1969, the Selfoc^{TM} lens, a cylindrical gradient-index (GRIN) lens was demonstrated [1]. The Selfoc lens is a trade mark of Nippon Sheet Glass Co., Ltd and GRIN is the general name that comes from Gradient Index. The

Fig. 1. Selfoc lenses.

Selfoc lens is the cylindrical rod that has the lens effect in the axial direction. Figure 1 is a picture of the Selfoc lens. The refractive index is the highest on the center axis and gradually deceases in the radial direction. The refractive index of the Selfoc lens is described by

$$N(r) = n_0 \left(1 - \frac{A}{2}r^2\right), \tag{1}$$

where r is the distance from the center axis and n_0 is the refractive index on the center axis and A is the coefficient that describes the decrease of the refractive index of the Selfoc lens.

When the refractive index of the Selfoc lens is described by Eq. (1), the ray trajectory of in the Selfoc lens is described by the combination of the sine function and the cosine function:

$$r = r_0 \cos(\sqrt{A}z) + \dot{r}_0 \sin(\sqrt{A}z). \tag{2a}$$

Also, the angle of the ray trajectory is described by

$$\dot{r} = -r_0\sqrt{A} \sin(\sqrt{A}z) + \dot{r}_0\sqrt{A} \cos(\sqrt{A}z), \tag{2b}$$

where dot denotes the differentiation in the optical axis, which means the angle of the light ray. The pitch of the ray is related to square root of A, which describes the decrease of the refractive index. When the product of the square root of A and distance in the optical axis z equals 2π, the ray trajectory makes one pitch of the sinusoidal curve:

$$\sqrt{A}z = 2\pi. \tag{3}$$

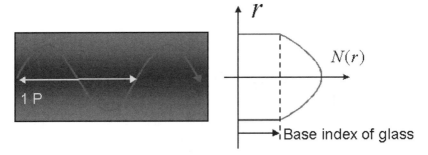

Fig. 2. Refractive-index distribution of Selfoc lens and ray trajectory.

The concept of refractive-index distribution and ray trajectory is shown in Figure 2.

When the diameter of the rod is the order of millimeters, the Selfoc lens can be used as a convenient microlens. It was also studied when the diameter of the rod is a few tens to $100\,\mu m$; the rod works as a multimode optical fiber with a very small modal dispersion [2]. The optical fiber was made by drawing the Selfoc lens in a small diameter that is called Selfoc fiber. The first commercial optical fiber transmission system was installed in the Florida Disney World in United States in 1978 using the Selfoc fiber.

3 Application of the Selfoc Lens

The dense wavelength division multiplexing (DWDM) system is the significantly important telecommunication system with a huge capacity. The DWDM system is multiplexing optical signals with different wavelength channels. The wavelength band used for the DWDM is selected for the wavelength range where the EDFA (Er-doped fiber amplifier is used: around 1530-1625 nm). The Selfoc lens has the important role in multiplexing and demultiplexing the different wavelength signals. Figure 3 shows the schematics of the multiplexer using the Selfoc lens. A light signal of wavelength $\lambda 1$ is incident on the optical fiber of port 1. The light from the port 1 fiber is collimated by the Selfoc lens and reaches the optical filter. This filter reflects the light of the $\lambda 1$ to the Selfoc lens. The reflected light is focused into the output fiber. The light signal of $\lambda 2$ is incident on the optical fiber of port 2. The light from port 2 is collimated by another Selfoc lens to reach on the optical filter from the right. The light of wavelength $\lambda 2$ passes through the filter and is incident on the Selfoc lens. The light is focused on the output fiber and multiplexed with a light of wavelength $\lambda 1$. This process can be cascaded to multiplex the lights of the other wavelengths: $\lambda 3$, $\lambda 4$, $\lambda 5$, ..., and λn, as shown in Figure 3. Usually, this optical filter is made from a dielectric multilayer.

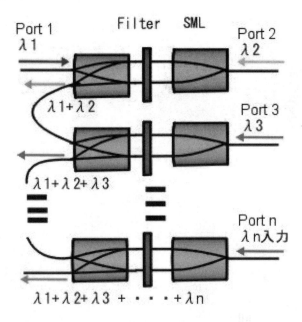

Fig. 3. Schematics of the wavelength multiplexer using the Selfoc lens.

On the other hand, if we reverse the direction of the arrows, lights of different wavelengths $\lambda 1$, $\lambda 2$, $\lambda 3$, ..., and λn are incident on the input fiber. The lights are demultiplexed by the filer and come out of the corresponding output port.

This multiplexer/demultiplexer with a cascaded structure is made from three port modules that consist of a pair of Selfoc lenses, fibers, and filters. This most fundamental structure is used not only as a multiplexer for signal lights but also as a multiplexer for the pumping light and signal light for EDFA and polarization combiners.

4 Vertical Integration

One of the greatest advantages of the Selfoc lens is that we can use flat surfaces. We can coat a dielectric thin-film filter directly on the surface of the Selfoc lens. Figure 4 shows the filter on the Selfoc lens. The filter consists of about 80 dielectric layers. Using the filter on the Selfoc lens, we can make the optics of the multiplexer simpler than shown in Figure 4. Since the filter is coated on the lens surface, we can bring the lenses in direct contact with each other. By eliminating the distance between lenses, we can make the ideal telecentric optical system. Since the telecentric optical system makes the chief ray parallel to the optical axis, the assembly of the multiplexer become easier and will result in the cost reduction of the fabrication process.

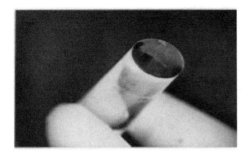

(a) Appearance   (b) Layer structure

Fig. 4. Filter on the Selfoc lens.

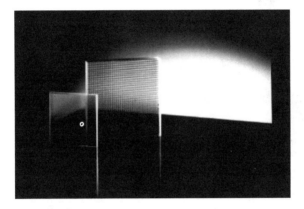

Fig. 5. Appearance of the planar microlens.

5 Planar Microlens: Horizontal Integration

In addition to the DWDM technology, it is also important to treat the optical signal in parallel. A planar microlens is a two-dimensional (2-D) integrated microlens lens array that brings about many possibilities of parallel optics. Figure 5 shows the appearance of the planar microlens.

The planar microlens is fabricated by the selective ion-exchange technology [3], as shown in Figure 6. First, we prepare the planar glass substrate. Second, metal film that prevents the ion exchange is evaporated on the substrate. Small windows are opened on the metal mask. Third, the masked substrate is immersed into the molten salt for the ion exchange. Special ions are diffused into the substrate selectively through the windows on the substrate. Since the ion has larger diameter and larger electron polarizability, the ion makes swells on the surface and a larger refractive index is obtained, as shown in Figure 6.

By using the swelled structure, we can have a large-NA (numerical aperture) lens that can be mainly used for light coupling between the LD and

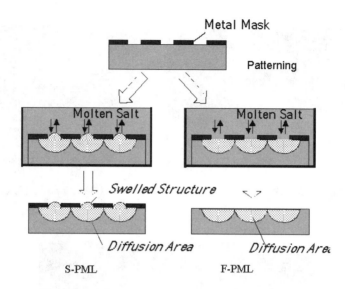

Fig. 6. Fabrication process of the planar microlens.

Table 1. Fundamental Characteristics of PML

	S-PML	F-PML
Lens pitch	250 μm	250 μm
Lens diameter	85 μm	250 μm
Back focal length	78 μm (at 1300 nm)	700 μm (at 1550 nm)
NA	0.48	0.18

optical fiber. On the other hand, if we polish the surface of the planar mi-crolens, we can have a microlens with a flat surface. The lens effect comes from the refractive-index distribution. The flat lens has a moderate NA around 0.2, which is suitable to accept the light from optical fiber to be collimated. We introduced the acronym S-PML to denote the swelled structure and F-PML for the lens with flat surfaces. Fundamental characteristics of the planar mi-crolens are summarized in Table 1. We are preparing S-PMLs and F-PMLs with pitch (the center-to-center space between each lens) of 250 μm. In many cases, since arrayed optical elements such as the LD array, PD array and op-tical fiber array are made with a pitch of 250 μm, the standard PML with a pitch of 250 μm is convenient for making the arrayed module.

The S-PML is used for the parallel interconnection, as shown in Figure 7. The light signal from the LD array is focused into the tape fiber array and transmitted to the PD array. The S-PML is used for the following reasons; the (1) S-PML is used for the window of the LD and PD array. Since it is important to avoid moisture or humidity to keep the reliability of the LD and PD array, the window that can keep the hermetical sealing is important. The S-PML

S-PML

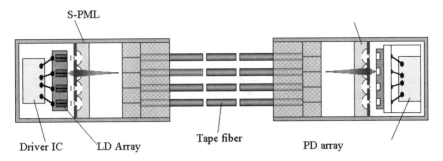

Driver IC   LD Array   Tape fiber   PD array

Fig. 7. Application for a parallel interconnection.

that is made from a glass is suitable for the window for hermetical sealing. We prepared the S-PML adhered on a metal holder using low-temperature-melt glass. Equation (2) is used to obtain sufficient light-coupling efficiency and tolerance between a LD array and an optical fiber array.

Figure 8 shows the statistical data of the coupling loss between LDs and single-mode optical fibers. The average coupling loss was as small as 4.28 dB. Since the parallel optical interconnection uses a very short reach connection, the coupling loss is not significantly important; direct coupling between a LD array and an optical fiber array can be an alternative solution. The other advantage of using the S-PML for a LD and an optical fiber coupling is the relaxing of the tolerance of the optical alignment. In the case of direct coupling, the insertion loss increases sensitively, corresponding to the distance between the LD and optical fiber. When we use the S-PML, the tolerance of the distance between the S-PML and optical fiber is around ± 100 µm.

F-PML is made when we polish the surface of the planar microlens after the ion exchange. The lens effect comes from the internal index distribution. One of the most fundamental applications of the F-PML is the light collimation of an optical fiber array. A fiber collimator array can be made by locating the fiber near the focal point of the each lens of the F-PML. Figure 9 shows the collimator array. The collimator array is usually used as a pair. The lights from one fiber array is collimated by the lens array and focused again into another fiber array by another lens array. Figure 9 also shows the light-coupling efficiency of pair of the 8-channel collimator array. The average coupling loss is as small as 0.5 dB. Between the lens arrays, since each light is collimated, we can insert optical elements such as filters, mirrors, and so forth to make functional devices. Since the light emerges after the lens is spread by the diffraction, the collimation length (working distance) between lens arrays is limited. Since the diffraction angle is inversely proportional to the diameter of the lens, the working distance is limited by the lens diameter.

Figure 10 shows the theoretical limit of the working distance corresponding to the lens diameter. The lens diameter should be designed for the required working distance.

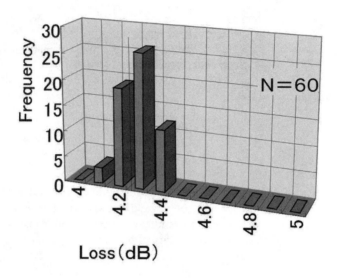

Fig. 8. Statistical data of coupling loss between LD and the single-mode fiber using PML.

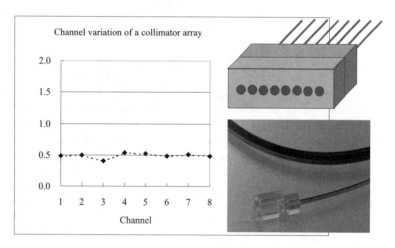

Fig. 9. Collimator array and light-coupling characteristics.

Figure 10 also shows the insertion loss of the collimator of the PML. One pair of collimators is prepared and light from one fiber is coupled into another fiber. The lens diameter of the PML is 250 μm. The insertion loss is increased when the working distance becomes larger than 10 mm.

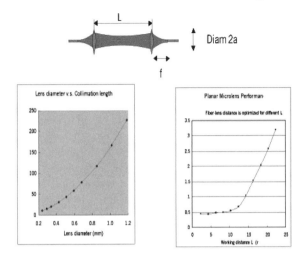

(a) Working distance vs. lens diam   (b) Light coupling loss vs. working distance

Fig. 10. Light-coupling loss of collimator: (a) working distance versus lens diameter (theoretical value); (b) light-coupling loss versus working distance (experimental value).

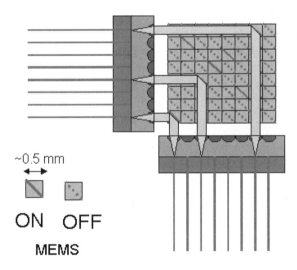

Fig. 11. Matrix switch using PML and MEMS.

We can make functional devices using the pair of the collimator array. Figure 11 shows the schematic of matrix switch using the collimator array and MEMS (microelectromechanical systems) micromirror. The MEMS consists of an array of mirrors. The collimated light can be cross-connected to the desired fiber by changing the status of each mirror.

6 Summary

We have introduced the Selfoc lens and the planar microlens that are based on the ion-exchange technology. The Selfoc lens is widely used in the field of DWDM to make the wavelength multiplexer and demultiplexers. The planar microlens is a monolithically integrated array of microlenses that is used for optical interconnection. An optical communication system will evolve to aiming future optical cross-connection. Since these microlenses that are made by the ion-exchange technology have a wide flexibility, the lenses will provide the solution for creating fundamental optical systems.

References

1. Kitano, I., Koizumi, K., Matsumura, H., Uchida, T., and Furukawa, M., A light-focusing fiber guide prepared by ion-exchange techniques, Proc. 1st Conf. on Solid State Devices, Tokyo, 1969; supplement to J. Jpn Soc. Appl. Phys. 39, 63-70 (1970).
2. Gloge, D., Chinnock, E. L., and Koizum, K., Study of pulse distortion in Selfoc fibers, Ellectron. Lett. 3, 856-857 (1972).
3. Oikawa, M. and Iga, K., Distributed index planar microlens, Appl. Opt. 21(6), 1052-1056 (1982).

Planar-Integrated Free-Space Optics: From Components to Systems

Matthias Gruber and Jürgen Jahns

FernUniversität Hagen, Lehrgebiet Optische Nachrichtentechnik, Universitätsstr.
27, 58084 Hagen, Germany
matthias.gruber@fernuni-hagen.de

1 Introduction

Planar-integrated free-space optics (PIFSO) is an integration concept for classical free-space optics that was proposed in 1989 [1] and has since evolved into an open integration platform for micro-opto-electro-mechanical systems (MOEMS), sometimes also called optical MEMS. This chapter aims at providing an overview of this development that is characterized by a shift from component-related design and fabrication issues to system- and application-related ones. We will discuss the potential of PIFSO within the fast-growing field of microsystems engineering and compare it with that of alternative integration approaches.

1.1 The Trend Toward Microsystems Integration

The main driving force for the ongoing trend toward system integration in many technical disciplines is the huge economic benefit that can be expected judging from the example that integrated microelectronics has been setting during the past four decades since the invention of the integrated circuit (IC) [2]. The technological development in this field has been governed by Moore's law, which states that the integration density on very large scale integration (VLSI) chips will double approximately every 18 months [3]. This has lead to extremely powerful and compact ICs and to chips that contain entire electronic microsystems; for example, state-of-the-art microprocessor chips like Intel's Pentium IV or AMD's Athlon contain of the order of 10^8 transistors that are arranged in various functional subcircuits. Furthermore, the fabrication of such systems as a whole using lithography-based mass fabrication techniques [4,5] made it possible to keep the cost per piece comparatively low although the involved technology is meanwhile extremely sophisticated and requires billion-dollar investments.

It goes without saying that there is an immense economic interest that disciplines like mechanics or optics progress in the same direction. An important

a)   classical optical setup

b)   "stacked microoptics"

c)   "microoptical bench"

d)   PIFSO

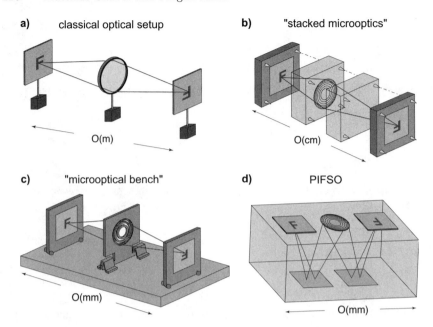

Fig. 1. Free-space optical integration concepts with their respective typical system dimensions.

condition put forward by the industry is that the approaches for microme-chanical and microoptical system integration be compatible with the estab-lished methods of VLSI electronics in order to reduce the cost for developing suitable design and fabrication processes and in order to make a convenient integration of electronic, optical, and mechanical functions on a common tech-nological platform possible. Adhering to this condition thus not only allows one to make many common-use technical systems smaller, better, and less expensive but to construct microtools and micromachines with entirely new applications – something that was merely a vision not too long ago [6].

In recent years, numerous approaches for microsystems integration have been proposed that are based on a variety of different fabrication methods and packaging concepts as well as specially engineered materials [7,8]. Arguably, the highest development level in the field of optics has been achieved so far with integrated waveguide components which were first proposed in 1969 [9] and are meanwhile ubiquitous in modern optical communication equipment. A prominent example is the arrayed waveguide grating (AWG) that is used for wavelength division multiplexing (DWDM) applications [10]. However, there are also integration approaches for classical free-space optics; the three most important ones are refered to as "stacked microoptics," "microoptical bench," and PIFSO. Figure 1 shows schematically how respective integrated systems would evolve from a very basic classical optical laboratory setup, namely an imaging system consisting of a lens, an input device, and an output device.

1.2 Free-Space Optical Integration Approaches

Stacked microoptics is the oldest one of the three integration approaches shown in Figure 1b and was proposed in 1982 by Iga et al. [11]. They suggested realizing optical components on planar substrates that are then stacked together to form the respective complete system. Components that are located in the same plane can thus be fabricated in parallel on the respective substrates, and alignment complexity is reduced from the six degrees of freedom (three translatory and three rotatory ones) for each discrete component in a classical laboratory setup to three (one rotatory and two translatory ones). For the assembly, low-cost passive alignment methods based on structures that are integrated into the planar substrates such as V-grooves can be employed. It is recommendable to realize the optical functionalities through gradient index (GRIN) elements because, then, the substrate surfaces can be kept flat, which simplifies the assembly and ensures that the system can be easily sealed and thus protected from disturbing environmental influences. The major shortcoming of the stacked microoptics approach is that the achievable level of integration is rather limited; a considerable amount of assembly and alignment work remains unavoidable.

The microoptical bench approach has its roots in micromechanics and, in particular, in the modern surface micromachinig techniques that have been developed over the past two decades [12]. They are based on the successive deposition of thin layers of different materials on a planar wafer (typically silicon) that can then selectively be removed by structuring and etching techniques very similar to those used in microelectronic mass fabrication. In this way, it is possible to fabricate free-standing cantilevers, beams, or bridges and even hinged structures that can be flapped upright and fixed in this position with appropriate latching mechnisms. Wu et al. [13] suggested applying this approach to construct miniature versions of classical optical bench setups by integrating the optical components into the hinged parts; the approach is schematically shown for a Fresnel zone plate in Figure 1c. The integration level that can thus be achieved is very high because most of the alignment is already done through the lithographic structuring that defines size and position of the hinged structures. Choosing the microoptical bench approach is particularly appealing when electromechanical actuators are to be integrated into a microsystem because they are fabricated with essentially the same techniques. However, there are also some weak points: One is robustness and the relative difficulty to protect microoptical bench systems from disturbing environmental influences. Another is the limited choice of materials and the fact that they are (at least up to now) often not optimal for implementing optical functionalities; for example, using silicon, a lens for the visible-wavelength range can only be implemented as (purely amplitude modulating) a Fresnel zone plate, which results in a very low efficiency. Finally, the surface-parallel orientation of the optical signal paths is certainly well suited when the integrated system incorporates optical fibers that are attached parallel to the substrate; however, it is not ideal for applications with I/O devices that are operated in a surface-normal manner such as optoelectronic chips with (arrays of) vertically emitting lasers or photodetectors. The surface-parallel approach is not recommendable either for applications that involve massively parallel communication.

Especially when large two-dimensional emitter, modulator, or detector arrays are to be used in a microsystem, PIFSO is the integration concept of choice because it perfectly matches the operational characteristics and the mounting requirements of these I/O devices. The idea of PIFSO is to miniaturize and "fold" a free-space optical setup into a transparent substrate of, typically, a few millimeters thickness in such a way that all optical components fall onto the plane-parallel surfaces. Passive components like lenses or diffraction gratings can then be integrated into the surface; for example, through surface relief structuring, active components like the aforementioned optoelectronic I/O devices can be bonded on top [14]. Thus, the substrate provides the mechanical support for the optical system and the bonded devices and may also be used for thermal purposes such as heat spreading and as a carrier for electrical wires. A reflective coating ensures that optical signals propagate along zigzag paths inside the substrate. Since all passive components are arranged in a planar geometry, the optical system can be fabricated as a whole using mask-based techniques. This approach provides lithographic precision concerning the positioning of components and high accuracy for the required relief structures because the well-established and sophisticated etching techniques of the semiconductor industry can be applied. Due to the integration into a rigid substrate, the passive optical system does not require further adjustment and it is well protected against disturbing environmental influences. From the topological point of view, the planar integration approach yields a fully three-dimensional (3-D) system architecture but requires only 2-D fabrication complexity. This allows one to implement highly parallel and complex (regular and irregular) optical interconnect schemes involving large fan-out, fan-in, and filtering operations. PIFSO modules are therefore particularly well suited for short-range communication applications [15], which typically require a level of parallelism that can hardly be provided by other technological approaches, including fiber optics.

A comparison of PIFSO with the stacked integration approach leads to the conclusion that both concepts share the use of planar substrates onto or into which optical components are integrated. However, PIFSO proceeds to a higher level of integration because the components of more than one functional plane of a classical optical laboratory setup are integrated into the same substrate surface. This saves much assembly and alignment work. The similarities between PIFSO and stacked microoptics suggest that it may be interesting to combine both concepts in the sense that PIFSO modules are used as components of a larger stacked microsystem. In the following section, we will narrow down the discussion of microsystems integration on PIFSO modules and address component-related issues.

2 Components for PIFSO-Type Microsystems

The components of a PIFSO-type microsystem can be divided into two groups: Passive optical components the functionality of which is set during fabrication and remains then fixed and active ones that can be reconfigured or controled during operation of the system. Active components are mostly based on electrooptic or electromechanic effects and are a major subject of general photonics research. We will cover them here only briefly by addressing some issues closely related to PIFSO and then proceed to passive optical components and their design.

2.1 Active Components

Since short-range communication is the most prominent field of applications for PIFSO modules, emitter, modulator, and detector devices that can be arranged in large 1-D and 2-D arrays are of primary interest. Among the emitters, this brings vertical-cavity surface-emitting lasers (VCSELs) to our attention. VCSELs were invented in 1979 at the Tokyo Institute of Technology [16] and have since experienced a rapid development. Their main characteristic is a short, vertically oriented laser cavity confined by two Bragg mirrors that lead to an emission of the generated radiation vertical to the chip surface. Various types of VCSEL structure (e.g., etched mesa, proton implanted, dielectric apertured, buried heterostructure) have been proposed, the details of which are described, for example, in Ref. [17]. VCSELs have favorable electrical and optical properties like low threshold current, high electrooptic conversion efficiency, axial single-mode operation, and a sysmmetric beam profile with a low numerical aperture. Particularly favorable for the use in PIFSO systems is the fact that the light is emitted orthogonally with respect to the chip surface, which is ideal when the VCSEL chip is bonded onto the transparent substrate. For the bonding, flip-chip techniques [18] can be applied that ensure micron-scale accuracy and low parasitic capacitances, an issue that is important for high-speed modulation. By applying appropriate PIFSO-compatible heat-spreading mechanisms [19, 20] and by using state-of-the-art VCSELs with optimized electrical characteristics [21], it is currently possible to pack VCSELs in large arrays with pitches of the order of $50\,\mu$m without running into thermal problems. This has important consequences for PIFSO-type communication systems and, in particular, for the way information is transmitted. It is currently possible to employ direct modulation schemes (i.e., to use VCSELs both as light sources and modulators).

The situation was still different a couple of years ago. Light generation had to be separated from modulation to enable densely packed arrays. From the mid-1980s to the mid-1990s considerable R&D efforts focused on VLSI-compatible electrooptic modulator arrays. Bell Labs researchers at that time developed a technology to hybridly integrate GaAs multi quantum well (MQW) p-i-n diodes [22] on conventional silicon complementary metal oxide semiconductor (Si-CMOS) circuits [23] to construct optical modulator arrays. This technology was also used to build devices with a more complex functionality that are often called smart-pixel arrays. Although hybrid modulator chips have meanwhile lost some importance, they are still used in photonic system demonstrators and they are very suitable for integration with PIFSO modules [24]. The operational principle of MQW p-i-n modulators is based on the quantum-confined Stark effect. It means that there is so-called excitonic absorption at photon energies between the smaller and the larger energy gaps in the MQW band structure. The absorption coefficient thereby depends strongly on the electric field across the quantum wells. This can be used to control the degree of absorption and, thus, electrically modulate an optical signal. The achievable bandwidth thereby reaches well into the gigahertz range and the necessary switching energy is low, which makes MQW p-i-n modulators very suitable for high-speed computing and communication applications.

What remains to be said is that MQW p-i-n diodes can also be used as photodetectors like conventional p-i-n diodes if they are operated with the appropriate electrical control circuitry. For detectors in general, there is a larger variety of technological options than for modulators and especially emitters because it is not necessary to use semiconductor materials with a direct band gap; various material compositions (binary, ternary, quaternary, heterostructure) are available to optimize the detector performance for the respective wavelength band of interest [25]. In particular, it is possible to integrate photodetectors with CMOS circuits on silicon VLSI chips. An interesting alternative to detectors based on p-n junctions are devices with Schottky barriers because they can conveniently be realized with two interdigitating comblike structures that are lithographically fabricated out of a thin metal layer on any semiconductor wafer. Such metal-semiconductor-metal (MSM) devices can have both a large sensitive area and a short response time (in the picosecond regime). This is unlike p-i-n diodes and can be readily exploited to alleviate the criteria for design and assembly that are required to obtain a certain tolerance level in a complex system.

In connection with silicon-based (opto)electronic components and circuits, it is worthwhile to mention a fabrication concept that dates back into the early days of VLSI electronics and that has gained new drive in recent years. It is called silicon-on-insulator (SOI) and means that electrically insulating wafer materials are used. The difficulty thereby is to deposit a thin silicon layer of sufficient quality for subsequent VLSI CMOS processing on such a wafer. This problem has meanwhile been solved. Using a novel approach called ultrathin silicon on sapphire (UTSi), the realization of integrated circuits with various functionalities has been reported [26]. UTSi components feature superior electrical isolation, which allows one to integrate digital and sensitive analog circuits on the same chip. What is most appealing for integration with PIFSO modules is the fact that the substrate material sapphire is transparent for essentially all wavelength bands that are relevant for optics and optical communication. Hence, it is possible to access optoelectronic components such as metal semiconductor metal (MSM) detectors with optical signal beams from the top and the bottom sides of the sapphire wafer, which provides an additional degree of freedom for the systems design. Furthermore, it is conceivable to integrate an entire microsystem, including electrical circuits and PIFSO-type interconnects, on just one substrate.

Of fundamental importance for communication systems are – in addition to emitters, modulators, and detectors – components for switching. The functionality of a switch is similar to that of a modulator, but different performance features have different priorities. For a modulator, a high temporal bandwidth is very important so that it can handle high data rates, whereas for a switch, low attenuation and a high on-off contrast are crucial. It is rather difficult to fulfill the latter two criteria well with most electrooptic modulator components; however, it is comparatively easy with electromechanic micromirror devices, provided that an appropriate surface treatment and coating are applied. By steering an optical signal beam into or out of the pupil of the optical system, very high contrast ratios can be achieved. Micromirror arrays that are fabricated by applying the kind of surface micromachinig techniques mentioned earlier are already in use for optical cross-connect switching nodes [27] and it is conceivable to integrate them also with PIFSO systems.

2.2 Passive Components

Our discussion of passive components will start from a general point of view before the focus is shifted to PIFSO-related issues. In general, an optical component can manipulate both the phase and the amplitude of an impinging wave. Because most applications, especially in the field of communications, require high energy efficiency, nonabsorbing components that affect only the phase are, by far, the most relevant ones and will exclusively be considered here. Phase-modulating free-space optical elements can be classified using a variety of different criteria. We like to adopt the scheme of Figure 2, which depicts how one of the most basic functions, beam focusing, can be implemented.

In the horizontal direction, the components are classified according to their physical principles of operation. Figure 2 list the two main options. The first one is to manipulate a light beam through an appropriately engineered refactive-index profile within the substrate material, the other one is based on surface profiles (i.e., suitably shaped boundaries between otherwise homogeneous materials). This second component type can be designed to operate in reflection or in transmission; in the latter case, materials with at least two different refractive indices need to be involved. It is possible to implement essentially the same types of optical function in both modes of operation; however, using reflective components, the same optical effect is achieved with a much shallower profile due to the dual passage of the light beams (cf. Eqs. (2) and (3) in Section 2.3).

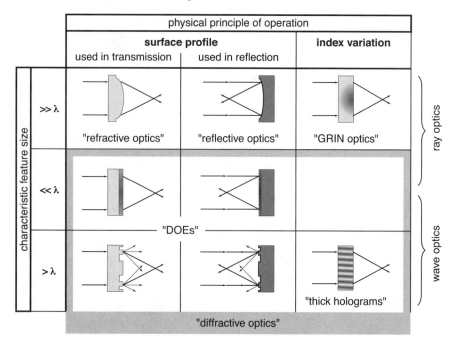

Fig. 2. Classification of passive free-space optical (micro)components.

The vertical classification criterion in Figure 2 is the characteristic feature size. Here, one can distinguish three categories referring to features much larger, much smaller, and of a similar size as the wavelegth λ of the light that impinges on the component. The feature size criterion decisively determines the mathematical effort that is necessary to describe the effect of the element on the electromagnetic wave sufficiently well. Components with smooth and slowly varying features can usually be modeled well using simple ray optics, which is based on a local plane-wave approximation. Their size may range from several tens of microns up to the macroscopic scale; hence, this category also covers (discrete) classical optical components. The three subdivisions in the top row of Figure 2 are often referred to as "refractive optics," "reflective optics," and "GRIN optics," respectively.

With features of the order of the wavelength, a ray-optical mathematical modeling is no longer possible. In this category, a wave-optical treatment is indispensable even though it is not, in general, necessary to apply a rigorous vector model of electromagnetic radiation; in many cases, including those that one usually encounters with PIFSO, a scalar thin-element approach as outlined in the next section is the best compromise between computational effort and physical accuracy. It is interesting and may be at first surprising that components of the second category with sub-λ features can also often be modeled with simple ray optics (at least when only the far field is of interest) even though the components of this and the previous category are considered to belong to the realm of diffractive optics. This is due to the fact that the optical wave cannot "resolve" the fine structures so that the optical effect is very similar to that of an (artificial) effective medium with a refractive index in between those of the involved physical media. It should, however, be noted that the effective refractive index is, in general, polarization dependent [28]. Diffraction effects do, of course, also influence the behavior of the components of the top category with the coarse structures; however, in this regime, diffraction can largely be considered as a limit for the resolution and can be ignored as far as the basic functionality of the element is concerned.

In principle, all of the element types shown in Figure 2 are compatible with the PIFSO approach; that is, it is conceivable to integrate them into a planar transparent substrate by processing the surfaces in a suitable way to obtain the desired relief profiles or by locally manipulating the material composition to generate the desired refractive-index distribution. For both tasks, there are design and fabrication methods that go along with established VLSI technology and batch processing. We want to mention here only a few examples from the plethora of techniques that have been developed over the last three decades (more detailed information can be found in Refs. [14, 29] and [30]).

The most important method to manipulate the refactive index of a material is ion exchange; thereby, a substrate is locally exposed to an environment that contains ions suitable to replace some of the ions of the substrate. The physical cause for the exchange is diffusion which depends on geometrical, chemical, and thermal parameters, as well as on an (externally applied) electrical field. These parameters allow one to tailor the optical functionality of a component very precisely [31]. It is possible to realize energetically efficient components with a comparatively high numerical aperture. Considerable limitations, however, exist in terms of the achievable index gradient. Fast variations of the phase delay as required, for example, for a fine diffraction grating can not be realized.

The surface-relief structure of an optical substrate can be modified in three ways: by locally removing material, by depositing new material, or through mass transport of the existing material. All three principles have been applied to fabricate (micro)optical components. The greatest practical importance so far have removal techniques and, in particular, etching techniques in conjunction with lithography. Most popular among the various possible combinations (binary or analog lithography, anisotropic or isotropic etching, etc.) are binary lithography together with a surface-normal etching technique such as reactive ion etching (RIE) or chemically assisted ion-beam etching (CAIBE). This approach can be applied sequentially with different mask patterns to obtain arbitrary staircaselike surface profiles [32]; the respective optical components are called multilevel surface-relief diffractive optical elements (DOEs).

After having been employed and refined for four decades by the semiconductor industry, binary lithography in combination with ion etching provides

234 Matthias Gruber and Jürgen Jahns

(at least currently) the highest accuracy of all material-removal techniques, especially when features of the order of the wavelength or below are to be formed. Therefore and because of the unsurpassed design freedom, multi-level DOEs have been used in most PIFSO system demonstrations. It should be mentioned that the multilevel approach also allows one to approximate continuous phase-delay profiles quite well provided that a sufficient number of discrete phase-delay levels are implemented. However, there is a trade-off between the phase resolution and the achievable phase gradient because no individual step of a staircase profile can fall short of the given lithographic minimum feature size. This and the increasing risk of alignment errors are the reason that the optimal DOEs are usually obtained with just two or three lithographic masks corresponding to four and eight discrete phase-delay levels, respectively.

2.3 Scalar Wave-Optical Modeling

As components that are fabricated with generically planar techniques, multi-level DOEs usually have a shallow surface relief of the order of the wavelength λ and, hence, can almost always be mathematically modeled as (infinitely) thin elements. In this model, which is also refered to as Kirchhoff approximation, a DOE acts as a simple multiplicative filter; that is,

$$U'(x,y) = U(x,y)\,F(x,y), \tag{1}$$

with $U(x,y)$ as the incoming and $U'(x,y)$ as the outgoing optical wave field; the complex-valued filter function $F(x,y) = A_F(x,y)\,\exp[-i\phi(x,y)]$ consists of a phase part $\phi(x,y)$ and an amplitude part $A_F(x,y)$. In the case of phase-only DOEs as envisaged here, A_F is constant. The phase distribution $\phi(x,y)$ is related to the surface profile $d(x,y)$ through

$$\phi(x,y) = \frac{2\pi}{\lambda}\,(n-1)\,d(x,y) \tag{2}$$

and

$$\phi(x,y) = \frac{4\pi}{\lambda}\,n\,d(x,y) \tag{3}$$

for the cases of transmissive and reflective operation, respectively.

In between the DOEs, light is assumed to propagate without restriction in a free-space optical setup (the effect of an interaction with the surrounding matter being solely expressed through its refractive index n). This can be adequately modeled by means of the Fresnel approximation of Kirchhoff's diffraction formula [33]

$$U(x,y,z_1) = \frac{e^{ik(z_1-z_0)}}{i\lambda(z_1-z_0)} \iint U(x',y',z_0)\exp\left(\frac{i\pi\left[(x-x')^2+(y-y')^2\right]}{\lambda(z_1-z_0)}\right)dx'dy',$$

which allows one to determine the light distribution $U(x, y, z_1)$ in a plane z_1 from the one in a plane z_0. By expressing the above Fresnel integral in the form

$$\iint \ldots = E(x, y) \iint E(x', y') \, U(x', y', z_0) \, e^{i2\pi(x'u + y'v)} \, dx' dy'$$

with

$$E(x, y) = \exp\left(\frac{i\pi(x^2 + y^2)}{\lambda(z_1 - z_0)}\right), \quad u = \frac{x}{\lambda(z_1 - z_0)}, \quad v = \frac{y}{\lambda(z_1 - z_0)},$$

it becomes obvious that light propagation through the free space can be implemented on a computer very efficiently by means of two multiplication operations and a fast Fourier transformation (FFT), which is a standard tool in commercial mathematics software packages:

$$U(x, y, z_1) \propto E(x, y) \, \text{FFT}\left\{ E(x', y') \, U(x', y', z_0) \right\} \tag{4}$$

With relations (1) and (4), the optical functionality of most PIFSO components and systems can be simulated adequately on a computer. Relation (4) is also the basis for Fourier-mathematical design techniques that play an important role for PIFSO.

2.4 Fourier-Optical Design Approach

Fourier-optical design techniques are based on the assumption that the light distribution $U'(x', y', z_0)$ just behind the DOE under consideration and the distribution $U(x, y, z_1)$ in a second plane of interest are Fourier-complementary, implying that the respective phase factors in expression (4) can be neglected (Fraunhofer approximation) or that they are compensated by lenses with a suitable focal length. The task is to find a Fourier pair $\{U', U\}$ that satisfies the given boundary conditions. Degrees of freedom for the designer arise from the fact that these boundary conditions do not, in general, determine U' and U completely in amplitude and phase. In practical applications, there is usually complete phase freedom, whereas both amplitude distributions are determined (to be constant in the space domain in order to obtain a phase-only element and to correspond to a certain desired intensity pattern $I(x, y)$ in the Fourier domain).

One of the most powerful approaches to solve this design task is called the "iterative Fourier transform algorithm" (IFTA). It is a numerical procedure that operates on sampled versions U'_{nm} and U_{nm} of the light distributions $U'(x', y', z_0)$ and $U(x, y, z_1)$, respectively, and it is particularly well suited when the design problem is discrete, as it is the case for periodic DOEs ("diffraction gratings" with distinct diffraction orders). Common to the many IFTA variants that have been developed over the past three decades [34–36]

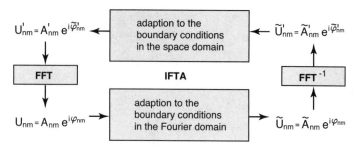

Fig. 3. Flowchart of the iterative Fourier transform algorithm.

is the basic flowchart of Figure 3. It depicts the gradual approach toward a solution through transformations between the two Fourier domains, followed each time by an adaption to the given boundary conditions of the respective domain. To ensure fast convergence of the IFTA, a variety of sophisticated adaption rules has been devised [36, 37]; since their description would go far beyond the scope of this discussion, we present only the most simple set of rules, which calls for a replacement of the respective actual amplitude distributions A_{nm} and $\tilde{A}'_{nm}$ after each Fourier transformation by the respective ideal amplitude distributions $\tilde{A}_{nm}$ and A'_{nm} while the respective phase distributions φ_{nm} and $\tilde{\varphi}'_{nm}$ are preserved:

$$\text{Fourier domain: } U_{nm} \rightarrow \tilde{U}_{nm} = \tilde{A}_{nm} \, \exp\left[i\varphi_{nm}\right] \propto \sqrt{I_{nm}} \, \exp\left[i\varphi_{nm}\right], \quad (5)$$

$$\text{Space domain: } \tilde{U}'_{nm} \rightarrow U'_{nm} = A'_{nm} \, \exp\left[i\tilde{\varphi}'_{nm}\right] \propto \, \exp\left[i\tilde{\varphi}'_{nm}\right]. \quad (6)$$

The IFTA can start at any corner of the loop and with any, even a random, light distribution, although it will converge faster and yield better results when started with a "good guess" for the element to be designed. The quality of the computed DOE is commonly judged by means of the diffraction efficiency η and the design error E. For η, we adopt the definition

$$\eta = \frac{\sum_{\text{des.}} |U_{nm}|^2}{\sum_{N,M} |U_{nm}|^2}, \quad (7)$$

where the summation in the numerator is carried out over the "desired" diffraction orders (i.e., the ones that should contain significant intensity contributions according to the design goal) and in the denominator over all $N \times M$ diffraction orders that are taken into account by IFTA. E quantifies the deviations from the desired intensity pattern I_{nm} according to

$$E = \frac{\sum_{\text{des.}} \left| |U_{nm}|^2 - I_{nm} \right|}{\sum_{\text{des.}} |U_{nm}|^2}. \quad (8)$$

It was proven mathematically that the adaption rules of Eqs. (5) and (6) will never lead to an increase of E after a complete pass of the loop [35]. Hence,

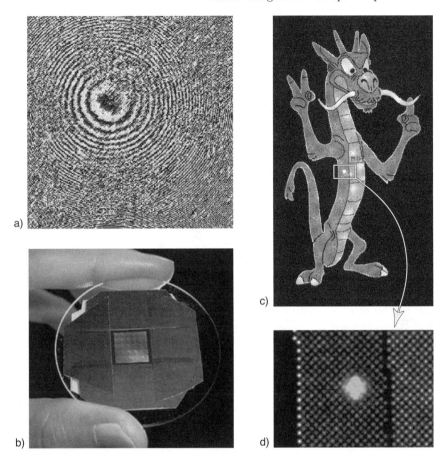

Fig. 4. IFTA design example: (a) computed phase delay profile (grey level coded); (b) eight-level surface-relief DOE on a fused silica substrate; (c) observed optical diffraction pattern; (d) blow-up of the central part of the diffraction pattern.

IFTA is a gradient descent method and will always converge; it may, however, stall in a local minimum of the error function. Nevertheless, with suitable adaption rules and starting conditions, it is possible to find for essentially all practical design problems a solution with $\eta > 90\%$ and $E < 1\%$, which is acceptable in most cases.

The particular strength of IFTA lies in the fact that it can provide (approximate) solutions for extremely complex problems and when an analytical solution is not known. An illustrative example is shown in Figure 4. Here, the design goal was to generate a DOE with a diffraction pattern that forms a cartoon-style image of a Chinese dragon [37]. After about 200 iterations, IFTA had produced the phase-delay profile of Figure 4a. A corresponding eight-level DOE in which this profile was periodically repeated 5×5 times

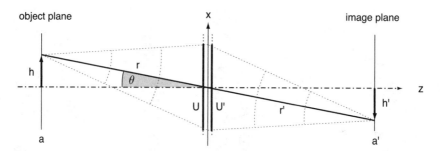

Fig. 5. Coordinate convention for the holographic design approach.

(Fig. 4b) generated the image of Figure 4c. It is composed of 500×500 focused diffraction orders and has excellent visual quality.

As far as PIFSO-type microsystems are concerned, a major application area for IFTA is the design of diffractive beam-splitter gratings for the purpose of implementing optical fan-out [38]. Due to the fact that usually a regular geometry of the spot pattern is needed, such DOEs are often called "array generator" or "array illuminator" (AI) gratings. For their realization as multilevel DOEs, an indirect design approach [39] may be applied, which means that, first, IFTA is used to compute a suitable phase-delay profile with unlimited phase resolution and then this profile is quantized to fit to the set of discrete phase-delay levels that can be physically implemented. It turns out that the phase quantization procedure can easily cause an unacceptable degradation of the optical performance of the AI grating in terms of the error E. To avoid this problem, special IFTA adaption rules and/or optimized quantization strategies should be applied [36, 40].

2.5 Nonrotationally Symmetric Lenses

Unlike the AI gratings mentioned earlier, many standard optical components for beam focusing or deflection do not require numerical design procedures like IFTA; they can, instead, be calculated analytically. The corresponding formulas are even comparatively simple if the geometry of the design problem exhibits rotational symmetry. This, however, is not the case in PIFSO systems. With a skew system axis, the analytical treatment is more complex and it leads to slightly different components to achieve an equivalent optical effect as in a rotationally symmetric system. A general mathematical formalism for "skew" optical systems can be derived on the basis of linear systems theory and canonical transformations between the skew system and a rotationally symmetric one [41, 42]. This approach, however, is rather lengthy and will not be further discussed here.

Instead, we want to outline a holographic approach and apply it to design optical components that are optimized for imaging in a skew geometry. The approach is called holographic because it resembles the recording procedure of a hologram and it is wellsuited for the design of planar DOEs. Essentially, it consists in an inversion of formula (1) in such a way that F is calculated from the two known light distributions U and U'. In holography, these would be the signal wave and the reference wave, respectively; for the design of an imaging lens, it is convenient to assume that U is a diverging spherical wave emerging from some point in the object plane, whereas U' is a spherical wave that converges toward the corresponding point in the image plane. The situation is graphically depicted in Figure 5 for the unfolded equivalent of a PIFSO setup where we assume (without loss of generality) that the inclination is in the x direction.

The two spherical waves can be mathematically expressed through

$$U(x,y) \propto \frac{1}{r(x,y)} \exp\left[i\,kr(x,y)\right] \qquad (9)$$

and

$$U'(x,y) \propto \frac{1}{r'(x,y)} \exp\left[-i\,kr'(x,y)\right], \qquad (10)$$

with $k = 2\pi/\lambda$ as the wave number and $r(x,y)$ and $r'(x,y)$ as distances from the two respective focal points. Let us now assume that the axial components of r and r' are much larger than their lateral components: $r_z \gg r_x, r_y$ and $r'_z \gg r'_x, r'_y$; in this so-called paraxial case, the square root expressions for r and r' can be approximated by the leading terms of a Taylor series expansion; that is,

$$r = \sqrt{r_x{}^2 + r_y{}^2 + r_z{}^2} = r_z + \frac{r_x{}^2 + r_y{}^2}{2r_z} - \frac{\left(r_x{}^2 + r_y{}^2\right)^2}{8r_z{}^3} + \cdots \qquad (11)$$

and an analogous relation for r'. The paraxial restriction furthermore allows one to neglect the amplitude variations of U and U' so that only the phase part of F has a significant dependence on x and y:

$$F(x,y) = \frac{U'(x,y)}{U(x,y)} \propto \exp(-i\,kr' - i\,kr) = \exp[-i\phi(x,y)] \qquad (12)$$

With the geometrical relations

$$r_x = x - h \approx x - \theta a, \qquad r_y = y, \qquad r_z = -a, \qquad (13)$$

$$r'_x = x - h' \approx x - \theta a', \qquad r'_y = y, \qquad r'_z = a', \qquad (14)$$

and only the two leading terms of the Taylor expansion (first-order approximation), one obtains from Eq. (12) the classical parabolic phase profile ϕ_1 for a lens:

$$\phi_1(x,y) = \frac{\pi}{\lambda f}(x^2 + y^2), \qquad \text{with} \qquad \frac{1}{f} = \frac{1}{a'} - \frac{1}{a}. \qquad (15)$$

Thereby, irrelevant constant terms have been omitted and the focal length f is defined via the object and image distances a and a', respectively, as indicated.

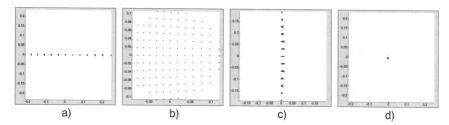

Fig. 6. Ray-tracing simulations of different lens types: (a–c) spot images at different z-positions around the nominal focus showing astigmatism with the parabolic phase delay profile of Eq. (15) in a skew imaging geometry; (d) stigmatic spot image with the adapted elliptic profile of Eq. (21).

Note that in the first-order approximation, there is no explicit dependence on θ, which means that a lens can basically image a whole plane onto another one.

The situation is different when one more term of the Taylor expansion is taken into consideration (third-order approximation). In this case, the resulting phase profile ϕ_3 contains three additional terms that are distinguished by the different dependence on variable θ:

$$\phi_3(x,y) = \phi_1(x,y) + \phi_{\text{spher.}} + \phi_{\text{coma}} + \phi_{\text{astig.}}, \tag{16}$$

with

$$\phi_{\text{spher.}} = \frac{\pi}{4\lambda}\left(\frac{1}{a^3} - \frac{1}{a'^3}\right)\left(x^4 + 2x^2y^2 + y^4\right), \tag{17}$$

$$\phi_{\text{coma}} = \frac{\pi}{\lambda}\left(\frac{1}{a'^2} - \frac{1}{a^2}\right)\left(x^3 + xy^2\right)\theta, \tag{18}$$

$$\phi_{\text{astig.}} = \frac{\pi}{2\lambda}\left(\frac{1}{a} - \frac{1}{a'}\right)\left(3x^2 + y^2\right)\theta^2. \tag{19}$$

These terms are responsible for spherical aberration, coma, and astigmatism according to the classical Seidel theory of optical aberrations [43]. Their relevance for PIFSO increases from top to bottom due to the increasing influence of θ and the fact that θ differs significantly from zero in a PIFSO system. In particular, astigmatism makes it impossible in such a system to obtain sharp image spots with the parabolic lens profile of Eq. (15). This is illustrated in Figures 6a–6c, which depict a z-scan through the (nominal) focal region; the spot images were obtained through ray-tracing simulations.

To reduce the aberration problem in PIFSO systems, it is recommended to employ lens profiles that correct (at least) astigmatism for the direction θ_0 of the skew system axis. Such a lens profile consists of two phase terms,

$$\phi_{\text{lens}}(x,y) = \phi_1(x,y) + \phi_{\text{astig.}}, \tag{20}$$

and can be expressed in the form

$$\phi_{\text{lens}}(x,y) = \frac{\pi}{\lambda} \left(\frac{x^2}{f_x} + \frac{y^2}{f_y} \right), \tag{21}$$

with

$$\frac{1}{f_x} = \left(\frac{1}{a'} - \frac{1}{a} \right) \cos^3 \theta_0, \tag{22}$$

$$\frac{1}{f_y} = \left(\frac{1}{a'} - \frac{1}{a} \right) \cos \theta_0. \tag{23}$$

An optimized lens for PIFSO is thus characterized by a slightly different focal length f_x versus f_y for the x and the y directions; as a consequence, points of equal phase (which determine the patterns for lithographic masks) now form elliptic curves. Ray-tracing simulations confirm that such a lens produces sharp spots in a skew system (Fig. 6d).

We want to conclude this section with the remark that it was sufficient in PIFSO demonstrators that have been built so far to compensate only astigmatism and neglect the other aberration types that depend more strongly on x and y and hence the numerical aperture of the imaging system. With a different fabrication technology that permits larger numerical apertures, however, it may become desirable or even necessary to compensate also coma and spherical aberration. The latter problem can be conveniently solved at the component level because $\phi_{\text{spher.}}$ does not depend on θ; coma, on the other hand, does depend on θ and may require a compensation at the system level.

3 System-Level Aspects of the PIFSO Concept

At the system level, the search for an optimal design is much more application dependent than at the component level. Nevertheless, there are some general recommendations that a system designer can follow. The related aspects of the PIFSO integration concept will be addressed in the first part of this section. In the second part, we will then exemplarily discuss the design and report about the performance of a real PIFSO system, namely a vector-matrix-type optical interconnect module.

3.1 Imaging with Arbitrary Magnification

One of the fundamental questions about the PIFSO approach is whether it is possible to design an equivalent planar-integrated version of any classical optical system despite the inherent geometrical restrictions of the integration concept. As far as imaging setups are concerned, this issue comes down to the more specific question of whether arbitrary magnifications can be implemented. Obviously, this is not the case for a single-lens system; here, the magnification is completely determined by the object and the image distance,

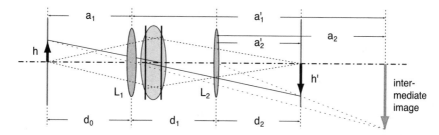

Fig. 7. Imaging with arbitrary magnification under the geometrical restictions of the PIFSO integration concept.

which are quantized if the object plane, image plane, and lens are to coincide with a substrate surface. In other words, a planar-integrated single-lens approach does not provide enough degrees of freedom.

To extend the design scope, it is common practice in optical system design to increase the number of elements in the system. We will show now that two lenses are, in principle, sufficient to build a PIFSO-type imaging system with arbitrary magnification. For clarity, the discussion is based on an unfolded geometry as shown in Figure 7; the PIFSO-inherent restrictions are thereby taken into account through fixed distances d_0, d_1, and d_2. Figure 7 depicts the intermediate image that is generated by the first lens L_1 and acts then as the object for the second imaging operation by lens L_2. From the geometrical relations

$$a_1 = -d_0, \qquad a_2 = a_1' - d_1, \qquad a_2' = d_2, \qquad (24)$$

and the basic relations between the two focal lengths f_1 and f_2 and the partial magnifications m_1 and m_2 [44] as well as their relation with the total magnification $m \equiv h'/h$,

$$a_1 = f_1 \left(\frac{1}{m_1} - 1 \right), \qquad a_2' = f_2 \left(1 - m_2 \right), \qquad m = m_1 m_2, \qquad (25)$$

one can, after some algebraic transformations, obtain two formulas that express f_1 and f_2 as functions of the given geometry and an (arbitrary) magnification m:

$$f_1 = \frac{m d_0 d_1}{m d_0 + m d_1 + d_2}, \qquad (26)$$

$$f_2 = \frac{d_1 d_2}{m d_0 + d_1 + d_2}. \qquad (27)$$

This solution of the magnification problem can be interpreted as follows: In first-order approximation, the two imaging operations by lenses L_1 and L_2 can be described as linear operations. Their combination is also a linear operation and can, therefore, be associated with an imaginary (thick) lens, the position of which depends on the ratio of f_1 and f_2. The multi lens system is thus

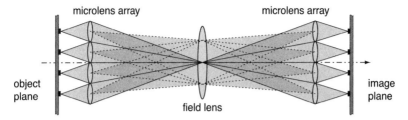

Fig. 8. Schematic of a hybrid imaging system.

reduced to an imaginary single-lens system that is not affected by geometrical restrictions.

3.2 Functional Efficiency and Fault Tolerance

In addition to the question of functional equivalence with classical systems, two other issues are important at the PIFSO system level: functional efficiency and fault tolerance. Optimizing functional efficiency means implementing a certain system functionality and achieving a certain system performance with the lowest possible system complexity and technological effort. Tolerant design means that this functionality is ensured even if system parameters deviate within a certain tolerance range from their ideal values. The two issues are often correlated because lower complexity and technological effort usually means fewer parameters that can vary and, therefore, a lower risk of accumulating errors that cause malfunction of the system.

An example for optimizing the functional efficiency is the so-called "hybrid" imaging approach [45]. It is shown in Figure 8 and recommended when only certain (regularly arranged) regions of interest in the object plane are to be imaged ("dilute array"). This is often the case when optoelectronic I/O devices are to be optically interconnected and, therefore, it is of high practical relevance for PIFSO modules. A hybrid imaging system comprises two microlens arrays that are imaged onto each other by means of a (global) field lens. The microlenses operate locally in the sense that they image only their respective regions of interest. The hybrid approach thus provides a large number of free-design parameters – the phase profile of each microlens can be individually optimzed for the respective imaging geometry – and it matches the space bandwidth product of the entire system with that of the problem; in other words, a hybrid imaging system provides exactly the number of spatial communication channels that are needed. This is unlike a conventional imaging system which would have to possess vast excess communication capacity in order to be able to resolve the regions of interest in the dilute array. Design and fabrication of such a system would, therefore, have to be significantly more complex. Experimentally a PIFSO-type hybrid imaging system with 2500 optical communication channels at a density of $400/mm^2$ was demonstrated [46]

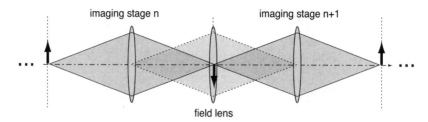

Fig. 9. Schematic of a fault-tolerant lens waveguide system.

which proves the suitability of this approach for massively parallel optical interconnect applications.

An example for improving the fault tolerance is the use of the so-called "lens waveguide" design approach. It has been applied for PIFSO-type optical interconnects that involve many zigzag steps between emitter and detector [47]. Figure 9 shows the basic idea: A communication channel is realized as a concatenation of many (usually identical) imaging stages that are connected through field lenses. In the ideal case, the optical signal propagates along the system axis and is not influenced by the field lenses; however, if the signal deviates from the axial propagation direction due to some perturbation, the field lenses "pull" it back and cause an oscillation around the ideal position. This stabilization effect is analogous to the one in a confocal laser resonator. Compared to a simple single-stage imaging system in which signals would increasingly walk off course, the waveguide-type interconnect system is thus much more fault tolerant. This has been confirmed in computer simulations and practical experiments about the effect of geometrical variations of the PIFSO substrate (thickness, wedge, temperature expansion) and variations of system parameters such as wavelength.

3.3 Fiber-PIFSO Optical Interface

It has already been mentioned that the PIFSO concept has a number of technological features that make it a prime option for short-range optical communication. To become commercially viable, however, it also has to be made compatible with fiber optics, which is the established technological platform of the long- and medium-range communication domain. For that purpose, a fiber-PIFSO optical interface was developed that provides a convenient way to plug fiber ribbons onto a PIFSO module to exchange optical signals between the two [48].

On the "fiber side," the interface was designed for MT-type ferrules, a specimen of which is shown in Figure 10a. It can house up to 12 fibers (single mode or multimode) that are arranged in a linear array with a 250-μm pitch. Two precision-machined holes with a 0.7-mm diameter serve as passive alignment structures. In conventional fiber-optical applications, two such ferrules would be connected with matching steel pins (cf. Fig. 10a). The fiber-PIFSO

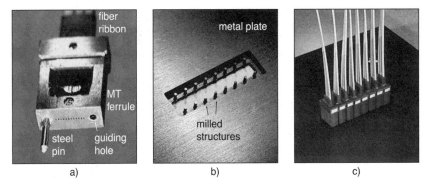

Fig. 10. (a) MT ferrule, (b) alignment pins milled out of a metal plate, and (c) MT ferrules atached to plate.

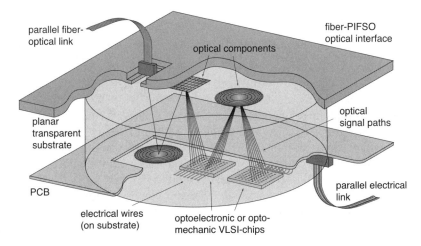

Fig. 11. Schematic illustrating the experimental state of the art concerning integration level and system complexity of PIFSO modules.

interface consists in a metal plate where similar (arrays of) pins have been milled out of the bulk material (Fig. 10b) in such a way that MT ferrules can be plugged onto them in a tight packaging scheme (Fig. 10c).

After alignment and mechanical fixation of the metal plate with the planar optical substrate on the opposite side, the fiber-PIFSO interface is ready for use as shown schematically in Figure 11. Experimental tests confirmed that this approach provides a positioning accuracy of the order of a few microns, which is sufficient for many practical applications. In particular, it is possible to relay optical signals reliably from fibers onto 20-µm × 20-µm targets on the planar substrate (Fig. 12). This target size is typical for optical detectors on optoelectronic VLSI circuits.

Fig. 12. Experimental performance of the fiber-PIFSO interface: Relay of optical signals from two (a) multimode and (b) single-mode fiber ribbons onto the same set of test targets on the planar optical substrate. The spot separation that is visible in (b) illustrates the typical positioning (in)accuracy.

In addition to the fiber-PIFSO interface, Figure 11 also illustrates the integration level and the system complexity that have so far been demonstrated experimentally with PIFSO-type modules. It includes the flip-chip bonding of optoelectronic VLSI chips onto the planar optical substrate and the electrical connection of these chips to conventional electrical printed-circuit boards.

3.4 Application: Vector-Matrix-Type Optical Interconnect Module

In this final subsection, we want to move from the rather general and fundamental discussion of PIFSO to more specific issues and look at applications. In addition to communication and information technology, the planar-integration concept is certainly also of interest for biology [49] and basic physics [50]. Many experiments in these disciplines involve optics and the respective setups could be made smaller and easily transportable ("lab on a chip"), which opens up entirely new technical and analytical possibilities. Practical demonstrations of PIFSO-type microsystems have so far mainly focused on information technology. Examples are modules for optical correlation [51] and pattern recognition [52], for optical clocking [53], and for optical information processing [54].

The example we want to discuss here in more detail is a planar-integrated multichip module (MCM) with vector-matrix-type (VM-type) optical interconnects [55]. VM-type architectures are ideal to demonstrate the advantages of PIFSO because their interconnect scheme is both of fundamental importance for optical switching (e.g., for cross-connects) and computing, and sufficiently complex so that it cannot so easily be implemented with standard electrical or waveguide-optical system approaches because these alternatives are essentially two dimensional.

Figure 13 shows the schematic setup of the multichip module. The VM-type interconnect architecture involves a 1-D optical fan-out stage followed by a 90°-rotated 1-D optical fan-in stage; these two stages are implemented along the x axis of the system. An additional optical fan-out stage along the y axis is used in this demonstration to optically program the central modulator chip that represents the matrix; the chip contains a 2-D array of MQW $p\text{-}i\text{-}n$ diodes in hybrid technology, as outlined in Section 2.1. The 1-D arrays of VCSELs serve as signal sources, and for detection, a charge-coupled device

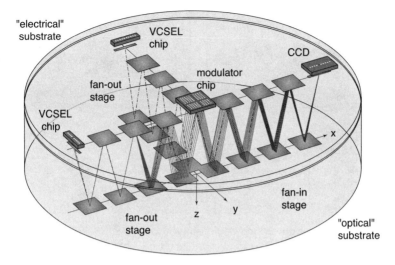

Fig. 13. Scheme of a PIFSO multichip module with VM-type optical interconnects.

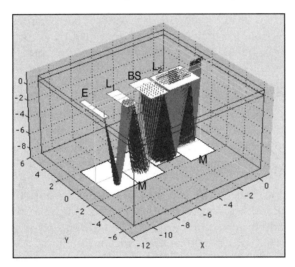

Fig. 14. Computer model of the fan-out stage of the multichip module. E, L_1, and L_2 are implement lenses, BS combines the functionality of a lens and a AI grating. M symbolizes a plane mirror.

(CCD) camera is used. All optoelectronic chips as well as the electrical lines for power and data supply (not shown in Fig. 13) are integrated on a 0.5-mm thin substrate, termed "electrical." All of the free-space optical components are integrated on a second 9-mm-thick substrate, termed "optical."

The design of the fan-out part of the optical system is based on multiple imaging using a lens assembly and an AI grating; the fan-in part is imple-

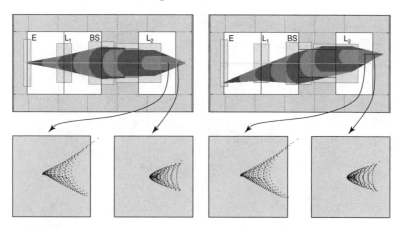

Fig. 15. Spot images from the image plane of the fan-out stage obtained through ray tracing simulations. The four squares in the lower part have a size of 20 μm × 20 μm and represent the respective targets on the modulator chip.

mented through an anamorphic imaging system that compresses the object field in one direction. To obtain good imaging quality, all parts of the optical system have to be adapted to the skew geometry. To this end, all lenses are modified according to Eq. (21) to reduce astigmatism. In addition to that, the lens assemblies are corrected for distortion at the system level. With these measures, one can expect good optical performance with precisely positioned spots and spot sizes of the order of 10 μm. This is confirmed by computer simulations on the basis of ray tracing. Figure 14 shows exemplarily a computer model of the fan-out part. Some of the corresponding spot images are depicted in Figure 15; apparently their structure is mainly determined by residual coma.

For the fabrication of the experimental demonstrator, the method described in Section 2.2 was used. Optical components were realized as multilevel DOEs and fabricated by means of binary lithography and reactive ion etching. Reflective coatings were realized with a thin layer of evaporated aluminum. Lithography in connection with a lift-off process was also used to realize the electrical lines on the "electrical" substrate. The optoelectronic chips were then flip-chip bonded onto their respective positions. Finally, the "optical" and the "electrical" substrates were aligned to each other using a specially developed optical method [56] and mechanically fixed.

Figure 16 shows the finished multichip module in an experiment where the basic functionality was tested by addressing individual VCSELs with electrical probe needles. In these tests, it was verified that all VCSELs could be electrically addressed and that the optical fan-out and fan-in worked as planned. Figure 17a shows the view with a CCD camera onto the 2-D modulator array. One can recognize the 20 × 10 target areas surrounded by the electrical lines. The 10 bright spots in the third line from the bottom are generated by the op-

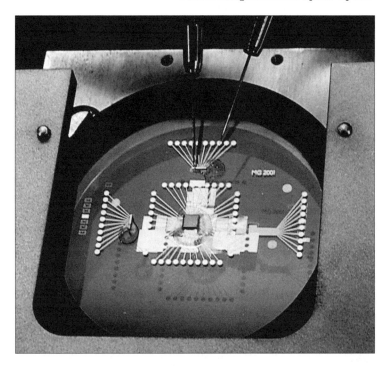

Fig. 16. Experimental test of the fully assembled multichip module with electrical probe needles.

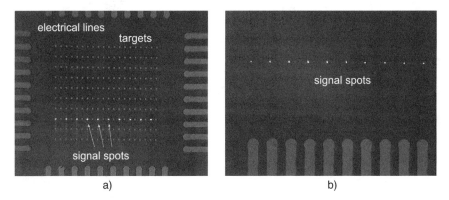

a)   b)

Fig. 17. Experimental demonstration of optical (a) fan-out and (b) fan-in the setup of Figure 16.

tical fan-out system when the third VCSEL is "on"; obviously, the respective targets are hit almost perfectly. It was found that the average spot diameter is 10.1 µm full width half-maximum, which is in good agreement with the theory and small enough to ensure good coupling efficiency. Figure 17b shows the view with a CCD camera onto the output area. Here, one can recognize 10 spots that result from the optical fan-in.

In conclusion, the planar-integrated multichip module with VM-type optical interconnects demonstrates the practical feasibility and the great application potential of the PIFSO integration approach.

References

1. Jahns, J. and Huang, A., Planar integration of free-space optical components, Appl. Opt. 28, 1602–1605 (1989).
2. Kilby, J. S., Invention of the integrated circuit, IEEE Trans. Electron. Devices ED-23, 648–654 (1976).
3. Moore, G. E., Cramming more components onto integrated circuits, Electronics 38, 114–117 (1965).
4. Sze, S. M., *VLSI Technology*, McGraw-Hill, New York (1987).
5. Moreau, W. M., *Semiconductor Lithography – Principles, Practices, and Materials*, Plenum Press, New York (1988).
6. Feynman, R. P., There's plenty of room at the bottom, *Engineering and Science*, California Institute of Technology, Pasadena, CA (1960).
7. Madou, M., *Fundamentals of Microfabrication*, CRC Press, Boca Raton, FL (1997).
8. Nalwa, H. S. (ed.), *Nanostructured Materials and Nanotechnology*, Academic Press, San Diego, CA (2002).
9. Miller, S. E., Integrated optics: An introduction, Bell Systems Tech. J. 48, 2059–2068 (1969).
10. Vellekoop, A. R. and Smit, M. K., Four-channel integrated-optic wavelength demultiplexer with weak polarization dependence, J. Lightwave Technol. 9, 310–314 (1991).
11. Iga, K., Oikawa, M., Misawa, S., Banno, J., and Kokubun, Y., Stacked planar optics: An application of the planar microlens, Appl. Opt. 21, 3456–3460 (1982).
12. Heuberger, A., *Mikromechanik: Mikrofertigung mit Methoden der Halbleitertechnologie*, Springer-Verlag, Berlin (1991) (in German).
13. Wu, M. C., Micromachining for optical and optoelectronic systems, Proc. IEEE 85, 1833–1856 (1997).
14. Sinzinger, S. and Jahns, J., *Microoptics*, Wiley-VCH, Weinheim (2003).
15. Jahns, J., Planar packaging of free-space optical interconnections, Proc. IEEE 82, 1623–1631 (1994).
16. Soda, H., Iga, K., Kitahara, C., and Suematsu, Y., GaInAsP/InP surface emitting injection lasers, Jpn. J. Appl. Phys. 18, 2329–2330 (1979).
17. Wilmsen, C. W. et al. (eds.), *Vertical-Cavity Surface-Emitting Lasers*, Cambridge University Press, Cambridge (1999).
18. Lau, J. H. (ed.), *Flip Chip Technologies*, McGraw-Hill, New York (1996).
19. Acklin, B. and Jahns, J., Packaging considerations for planar optical interconnection systems, Appl. Opt. 33, 1391–1397 (1994).
20. Gimkiewicz, C. and Jahns, J., Thermal management in planar optical systems with active components, Proc. SPIE 3226, 56–66 (1997).

21. Gulden, K.-H., Eitel, S., Hunziker, S., Vez, D., Gimkiewicz, C., Gale, M., and Moser, M., High density VCSEL arrays, Proc. 2002 IEEE/LEOS Annual Meeting, Glasgow, Scotland, 2002, pp. 129–130.

22. Miller, D. A. B., Chemla, D. S., Damen, T. S., Wood, T. H., Burrus, C. A., Gossard, A. C., and Wiegmann, W., The quantum well self-electrooptic effect device: Optoelectronic bistability and oscillation, and self-linearized modulation, IEEE J. Quantum Electron. 21, 1462–1476 (1985).

23. Goossen, K. W., Walker, J. A., D'Asaro, L. A., Tseng, B., Leibenguth, R., Kossives, D., Bacon, D. D., Dahringer, D., Chirovsky, L. M., Lentine, A. L., and Miller, D. A. B., GaAs MQW modulator integrated with silicon CMOS, IEEE Photon. Technol. Lett. 7, 360–362 (1995).

24. Fey, D., Erhard, W., Gruber, M., Jahns, J., Bartelt, H., Grimm. G., Hoppe, L., and Sinzinger, S., Optical interconnects for neural and reconfigurable VLSI architectures, Proc. IEEE 88, 838–848 (2000).

25. Wood, D. *Optoelectronic Semiconductor Devices*, Prentice-Hall, Englewood Cliffs, NJ (1994).

26. Kuznia, C. B., Flip chip bonded optoelectronic devices on ultra-thin silicon-on-sapphire for parallel optical links, OSA Technical Digest on Optics in Computing 2001, Lake Tahoe, Nevada, 2001, pp. 134–136 .

27. Kim, J., Papazian, A. R., Frahm, R. E., and Gates, J. V., Performance of large scale MEMS-based optical crossconnect switches, Proc. 2002 IEEE/LEOS Annual Meeting, Glasgow, Scotland, 2002, pp. 411–412.

28. Kuittinen, M., Turunen, J., and Vahimaa, P., Subwavelength-structured elements, in *Diffractive Optics for Industrial and Commercial Applications*, edited by Turunen, J., and Wyrowski, F., Akademie Verlag, Berlin (1997).

29. Herzig, H.-P. (ed.), *Micro-Optics*, Taylor & Francis, New York (1997).

30. Metev, S. M. and Veiko, V. P., *Laser Assisted Microtechnology*, Springer-Verlag, Berlin (1998).

31. Bähr, J. and Brenner, K.-H., Realization and optimization of planar refractive microlenses by Ag-Na ion exchange techniques, Appl. Opt. 35, 5102–5107 (1996).

32. Jahns, J. and Walker, S. J., Two-dimensional array of diffractive microlenses fabricated by thin film deposition, Appl. Opt. 29, 931–936, (1990).

33. Goodman, J. W., *Introduction to Fourier Optics*, McGraw-Hill, New York (1968).

34. Gerchberg, R. W. and Saxton, O. W., A practical algorithm for the determination of phase from image and diffraction plane pictures, Optik 35, 237–246 (1972).

35. Fienup, J. R., Phase retrieval algorithms: a comparison, Appl. Opt. 21, 2758–2769 (1982).

36. Turunen, J. and Wyrowski, F. (eds.), *Diffractive Optics for Indstrial and Commercial Applications*, Akademie Verlag, Berlin (1997).

37. Gruber, M., Diffractive optical elements as raster image generators, Appl. Opt. 40, 5830–5839 (2001).

38. Streibl, N., Beam shaping with optical array generators, J. Mod. Opt. 36, 1559–1573 (1989).

39. Mait, J. N., Understanding diffractive optic design in the scalar domain, J. Opt. Soc. Am. A 12, 2145–2158 (1995).

40. Gruber, M., Optimal suppression of quantization noise with pseudoperiodic multilevel phase gratings, Appl. Opt. 41, 3392–3403 (2002).

41. Testorf, M. and Jahns, J., Paraxial theory of planar integrated systems, J. Opt. Soc. Am. A 14, 1569–1575 (1997).
42. Testorf, M. and Jahns, J., Imaging properties of planar-integrated micro-optics, J. Opt. Soc. Am. A 16, 1175–1183 (1999).
43. Welford, W. T., *Aberrations of the Symmetrical Optical System*, Academic Press, London (1974).
44. Hecht, E., *Optics*, Addison-Wesley, San Francisco (2002).
45. Lohmann, A., Image formation of dilute arrays for optical information processing, Opt. Commun. 86, 365–370 (1991).
46. Sinzinger, S. and Jahns, J., Integrated micro-optical imaging system with a high interconnection capacity fabricated in planar optics, Appl. Opt. 36, 4729–4735 (1997).
47. Lunitz, B. and Jahns, J., Tolerant design of a planar-optical clock distribution system, Opt. Commun. 134, 281–288 (1997).
48. Gruber, M., ElJoudi, E., Sinzinger, S., and Jahns, J., Practical realization of massively parallel fiber-free-space optical interconnects, Appl. Opt. 40, 2902–2908 (2001).
49. Jahns, J. and Sinzinger, S., Microoptics for biomedical applications, Am. Biotechnol. Lab. 18, 52–54 (2000).
50. Birkl, G., Buchkremer, F., Dumke, R., and Ertmer, W., Atom optics with microfabricated optical elements, Opt. Commun. 191, 67–81 (2001).
51. Eckert, W., Arrizon, V., Sinzinger, S., and Jahns, J., Compact planar-integrated optical correlator for spatially incoherent signals, Appl. Opt. 39, 759–765 (2000).
52. Daria, V., Glückstad, J., Morgensen, P. C., Eriksen, R. L. and Sinzinger, S. Implementing the generalized phase-contrast method in a planar-integrated micro-optics platform, Opt. Lett. 27, 945–947 (2002).
53. Jahns, J., Gruber, M., Lunitz, B., and Stölzle, M., Optical interconnection and clocking using planar-integrated free-space optics, J. Opt. Soc. Korea 7, 1–6 (2003).
54. Gruber, M., Sinzinger, S., and Jahns, J., Planar-integrated optical vector-matrix multiplier, Appl. Opt. 39, 5367–5373 (2000).
55. Gruber, M., Multi-chip module with planar-integrated free-space optical vector-matrix-type interconnects, Appl. Opt. 43, 463–470 (2004).
56. Gruber, M., Hagedorn, D., and Eckert, W., Precise and simple optical alignment method for double-sided lithography, Appl. Opt. 40, 5052–5055 (2001).

Microoptical Structures for Semiconductor Light Sources

Werner Späth

Consultant for Osram-OS, Burgstallerstr. 10, 83607 Holzkirchen, Germany
W.F.Spaeth@t-online.de

1 Introduction

Semiconductor light sources like light-emitting diodes (LEDs), infrared emitting diodes (IREDs), and laser diodes have in the last years made a great step forward with regard to efficiency, reliability, and yield because of improvements in epitaxy, chip, and assembling technology. They cover the optical spectrum from blue up to the middle infrared and they are key components in the information and communication technology, the entertainment technology, in the automotive industry, and plenty of other applications. They work as simple indicator lamps, light sources for backlighting displays and sensors, transmitters in optical links, and many other functions. Their importance in the optical technologies will grow further in the future because of their high efficiency and low power consumption, small size and simple modulation of the intensity, low cost, and possibility of assembling them on printed-circuit (PC) boards together with electronic devices.

Here, we will concentrate on the technology and application of LEDs.

2 Semiconductor Technology

Light-emitting diodes (LEDs) are semiconductor diodes based on III–V semiconductors like the binary systems GaAs, GaP, and GaN, and the ternary and quaternary systems GaAsP, GaAlAs, InGaN, InGaAlAs, and InGaAlP and so on [1]. The band-gap and over-the-band-gap the color can be controlled by the combination and concentration of the III–V elements. Figure 1 gives an overview.

The correlation between band gap E_g (eV) and the wavelength λ is

$$\lambda(nm) = hc/E_g = 1240/E_g \,(\text{eV}). \tag{1}$$

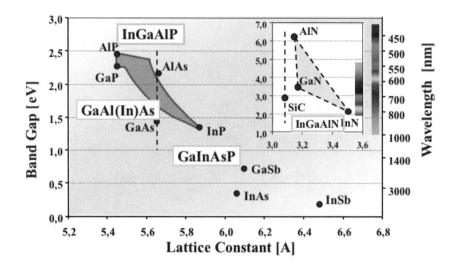

Fig. 1. III–V Technology.

Semiconductor lasers, the key components for fiber optics, were the big technological drivers for the development of the III–V semiconductor technology in the past. MOVPE (metal organic vapour-phase epitaxy) was developed mainly for laser fabrication. It allows the fabrication of quantum well structures with a great improvement in the performance. Using this technology for LED fabrication, a great progress in brightness and reliability for this devices could be achieved and made the fabrication of a GaN LED, the blue light-emitting diodes, possible. Now, LED in the colors red, green, and blue are available, the key colors for visualization. Combining a GaN LED with an appropriate phosphorus white light, other colors can be generated for illumination.

Figure 2 shows the fabrication steps of an LED chip. On a substrate, for yellow, amber, and red LEDs the substrate is GaAs, and for a blue and pure green LED, the substrate is SiC or sapphire. The epitaxial layers are deposited and the p-n junction is grown in situ, the p and n contacts are evaporated and photolithographically structured; then, the wafers are cut into chips. Standard chips have a surface area of 200^2–$300^2\,\mu m^2$ and are 200–$250\,\mu m$ thick. The front contact covers about $120^2\,\mu m^2$ of the light-emitting surface.

Figure 3 shows the vertical structure of an InGaAlP LED. The number, thickness, and composition of layers are comparable to that of a red edge-emitting laser. Thanks to this modern technology and the understanding of quantum well structures, the internal quantum efficiency of a InGaAlP LED for red, amber, and yellow light emission reached values in the range of $60\,\%$ to $80\,\%$, but only less than $10\,\%$ of the generated photons N are coupled out of the semiconductor. The basic process of generating a photon in the

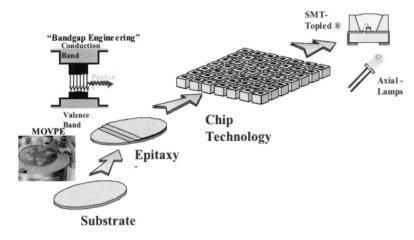

Fig. 2. Fabrication steps of a LED.

Epi-Structure

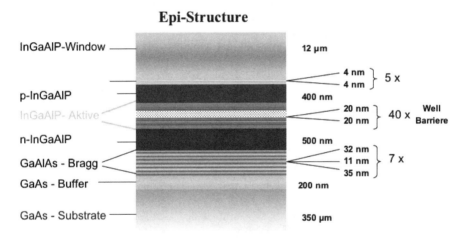

Fig. 3. InGaAlP technology for red, amber, orange, and yellow.

semiconductor is the radiant recombination of an electron and a hole, which are injected from the n side and p side, respectively. The direction in which the photon is emitted is statistical. This means, that this internal point source is omnidirectional. However, due to absorbtion at the back side of the LED and total internal reflection at the upper semiconductor/air interface most of the emitted photons are lost.

Using Bragg reflectors over the substrate consisting of $\lambda/4$-stack semiconductor similar to the technology used for vertical cavity surface emitting laser (VCSEL), the light output can be increased. However, because of the low difference of the refractive indices of the high and the low $\lambda/4$-semiconductor layers, only rays with a narrow angle to the perpendicular coupled out and contribute to the light extraction. One gains about a factor of 1.2 in comparison to the structure without Bragg reflector.

Only photons emitted into the cone defined by the angle of total reflection α can leave the chip over the plane surface of the LED. The others are reflected to the substrate. The angle of total internal reflection is

$$\alpha = \arcsin(1/n). \tag{2}$$

The high refractive index n of semiconductors – n lies between 2.5 for GaN and 3.4 for InGaAlP – causes a small angle of total reflection and, therefore, only a small percentage p of the generated photons N are coupled out over the upper surface (Fig. 4, left-hand side). The percentage p is proportional to the surface S

$$S = (1 - \cos(\alpha))/2 \tag{3}$$

which is cut out of the surface of the unit sphere by the cone with the half-angle α. The factor $1/2$ is valid if no Bragg reflector is used because then only photons directly radiated into the upper hemisphere contribute to p. Therefore, for p, we get the equation

$$p = S \times 100\%. \tag{4}$$

p reaches values in the range of 4.2 % for GaN and 2.2 % for InGaAlP in air and 11 % and 5 % in transparent plastics, for instance, epoxy with $n = 1.5$ if light leaves the semiconductor only through the upper surface.

If the nonabsorbing window layer (Fig. 3) is thick enough, a reasonable portion of photons are coupled out at the side walls of the LED too. In the ideal case, if the photon source is located on the axis of a cylindrically shaped chip and the thickness d of the window layer is the minimum $d = R^* \tan(\alpha)$ (R radius of the cylinder), the surface S° cut out of the unit sphere is

$$S^\circ = \sin(\alpha)/2. \tag{5}$$

This value is much higher than S. For practical applications, the photon source should be located inside the circle with radius $r = R/n$ in the radiant zone, where R is the radius of the cylinder and n is the refractive index of the semiconductor. This is the well-known Weierstrass condition. That means that in order to achieve a better coupling-out efficiency, the side walls should also be used for light extraction.

3 Microoptical Structures

The lateral dimensions of standard LED chips are 200–300 μm square. According to Eq. (5), the window layer should be 150–200 μm thick. The epitaxial deposition of such thick layers is principally possible, but very expensive. A practical approach is to bond a transparent wafer such as GaP on a InGaAlP

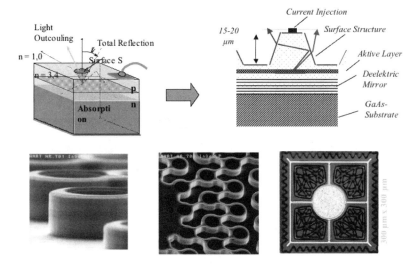

Fig. 4. Microoptical structures.

epi-wafer (wafer bonding) and then remove the absorbing substrate. With this technology, a very bright LED can be produced, but the bonding process is very pretentious and its quality has a great influence to the electrical and optical parameters of the diode.

Microoptical structures offer an another relatively simple solution to achieve a higher efficiency. Etching a pyramidal or cylindrical microstructure in the window layer, the coupling-out efficiency can be increased, but also other microoptical structures can be used [2]. A layer thickness of only 15–20 μm is necessary to achieve good coupling-out conditions . The diameter and the pitch of the micropyramids and cylinders have the same dimensions. Compared to the plane structure of the LED, the efficiency is increased by a factor of about 2. Figure 4 shows the plane and the microoptical structures of such a LED. The pyramidal structure guides the photons like an inverted taper. Reflected rays change their angle and therefore have a good chance of being coupled out at the opposite side wall or after a further reflection. In a cylindrical structure, only rays which are not totally reflected at the "first wall touch" leave the semiconductor. The reflected photons remain in the cylinder and will be absorbed.

4 Thin-Film LED

The LED with a microoptical structure on top still has the disadvantage that a large amount of the photons in spite of the Bragg mirror are lost by absorption in the substrate because only photon rays with a narrow angle perpendicular

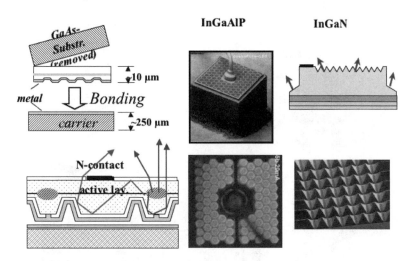

Fig. 5. Eutectic bonding.

to the chip surface are reflected and contribute to the efficiency, as already mentioned.

Replacing the dielectric mirror by a highly reflective metallic mirror, the losses on the back side can be reduced. The so-called thin-film LED uses this technology. A microoptical structure is etched in the eptaxial layer that is then evaporated with a highly reflective metal. The metal layer is a mirror and contact together. Then, the wafer with the metallized side eutectic is bonded to a carrier wafer. That is totally different from the already described wafer bonding process. Here, the highly reflective mirror replaces the transparent substrate. The advantage of this technology is that in addition to the lower costs, there are no restrictions on the chip size.

The efficiency, however, is very sensitive to the reflectivity of the mirror because multiple reflections are necessary to achieve high coupling-out values. The reflectivity can be varied by microoptical structuring of the top surface. If the microoptical structure consists only of a roughened surface, the optical output can be increased by a factor of 2 compared to a plane surface. The scattering process changes the direction of some of the internally reflected photons so that they can now leave the LED. This process can be further enhanced. If microoptical structures are used similar to them as shown in Figure 5, the factor can be increased to 3.

Blue-emitting diodes have a transparent substrate and, therefore, it has advantages to make the microoptical structure – in this case, a pyramidal structure – on the substrate. The coupling-out efficiency of such diode is nearly a factor of 4 higher compared with the former plane and p-side-up diode.

Figure 5 shows the principle bonding process and the surface structure of InGaAlP and InGaN diodes.

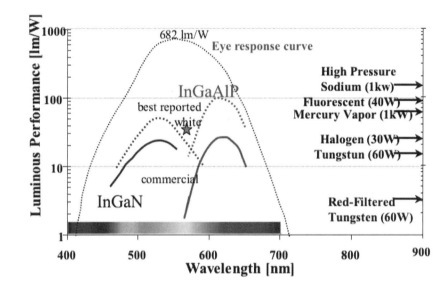

Fig. 6. InGaN and InGaAlP high-brightness LEDs.

5 Conclusion

Light-emitting diodes have a broad range of applications. Together with steady improvement of the basic materials and technology and the introduction of microoptical structures for coupling the light out of the semiconductor, the brightness of LED could be increased by more than a factor of 100 in the last 30 years and there is still potential for increase the efficiency. Compared to standard light sources, this is an impressive result (Fig. 6).

Acknowledgment

The author would like to thank N. Stath for providing the Figures.

References

1. Nakamura, S., Mukai, T., and Seoh, M., Candela-class high brightness InGaN/AlGaN double heterostructure blue light emitting diodes, Phys. Lett. 64, 1687 (1994).
2. Stath, N., and Härle, V., Wagner, J., The status and future development of innovative optoelectronic devices based on III-nitrides on Sc and on III-antimonide, Materi. Sci. Eng. B80, 224-231 (2001).

Microooptic Sensors

Albrecht Brandenburg

Fraunhofer-Institut für Physikalische Messtechnik, Heidenhofstrasse 8, D-79110
Freiburg, Germany
albrecht.brandenburg@ipm.fhg.de

1 Introduction

Optical sensing using microoptical devices is a promising area not only because
of the small dimensions and cost-efficient production technologies of those
devices. Many sensing principles are based on special features, which are only
provided by microoptics and integrated optics. The tip of an optical fiber, for
example, provides an extremely small detection area for point measurements.
The fiber coil and the integrated optical phase modulators are components,
essential for the operation of optical gyroscopes. The evanescent field of a
guided wave allows the construction of very sensitive chemical and biochemical
sensors. Often, very small sample volumes have to be analyzed in biochemical
analytics. Miniaturization of the optical detection principles, therefore, is an
important issue in analytics.

2 Interferometric Displacement Sensing

At the end of the 1980s, integrated optical interferometers had been developed intensively for the construction of rotation rate sensing and distance
measurement. One of the main obstacles for sensitive detection, which is, in
principle, possible with interferometers, is the undesired sensitivity to influences like temperature variations or mechanical vibrations. Also, a very high
stability of all components within the system is required. The periodic structure of the interferometer transmission curve is an additional problem, from
which an ambiguity of the detected values and the so-called signal fading
arise. Miniaturized sensors can be constructed in a way which makes the system insensitive to disturbing outer influences. The signal processing problem
is solved by using integrated optical phase modulators, by which – at least
over a certain range – linearized signals are provided.

For distance gauging, commonly the Michelson interferometer is used
whose basic principle is shown in Figure 1 [1]. One of the mirrors is mounted

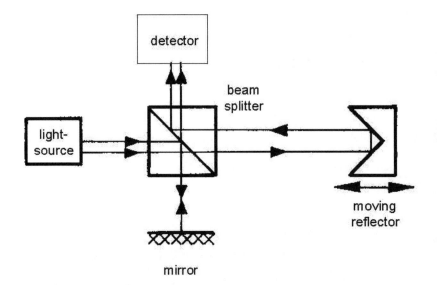

Fig. 1. Principle of the Michelson interferometer.

on the moved part; the other components are part of the measurement device. The transmission curve is tuned over one interference fringe when the mirror is moved over the distance of half a wavelength, resulting in the interferometer's high resolution.

The output intensity is proportional to the cosine of the phase difference $\Delta\varphi$:

$$I \sim 1 + \cos(\Delta\varphi). \tag{1}$$

The phase difference is basically given by the distance to be measured in units of the wavelength λ:

$$\Delta\varphi = 4\pi n z/\lambda. \tag{2}$$

A miniaturized version of this setup, as proposed in Refs. [2] and [3], is shown in Figure 2. The core element is an integrated optical (IO) device on silicon, which combines the function of the beam splitter, the reference mirror, a spatial filter, and a phase shifter. A semiconductor laser is used as a light source. The laser and the IO chip have to be temperature stabilized. One of the main issues is the stabilization of the laser wavelength.

The silicon chip is 7.5×7.5 mm; the dimensions of the sensor head are 74 mm in length and 17 mm in diameter. The measurement range is 700 mm at a resolution of <10 nm. The accuracy is mainly determined by laser drifts. The wavelength stability of the laser is 1×10^{-6}, resulting in an measurement accuracy of 100 nm for a working distance of 100 mm.

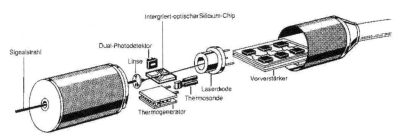

Fig. 2. Interferometric displacement sensor.

3 Laser Doppler Anemometry

Laser Doppler anemonetry (LDA) is a well-known principle for the contactless detection of the velocity of a moving element [1]. The scattered light from a moving surface is detected (Fig. 3). Light frequency is Doppler shifted. A difference frequency is observed when there are two light waves, one of them having a component in the direction of the moving object and the other one propagating in the opposite direction. This frequency is proportional to the velocity to be measured.

Another way to explain this effect is to consider the interference pattern produced by the two light waves on the moving surface. A scattering particle produces a modulated light intensity because of the spatial structure of the object illumination. A main problem of LDA is the detection of the direction of the movement. This can be provided by using two different light frequencies. This results in a moving interference pattern. Now, also at zero velocity, a periodic signal is found. The movement in one direction results in a decrease of the signal frequency and movement in the other direction results in an increasing frequency.

The incoming waves, oscillating at the frequencies ν_1 and ν_2,

$$E = E_0 \cos[2\pi(\nu_1 t)] \quad \text{and} \quad E = E_0 \cos[2\pi(\nu_2 t)], \qquad (3)$$

produce a frequency shifted signal, given by

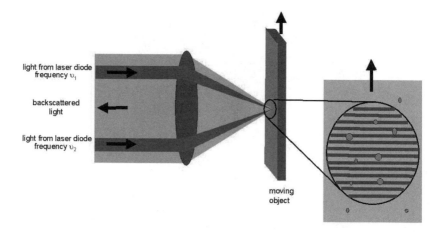

Fig. 3. Principle of laser Doppler anemometry.

$$I = \{E_0 \cos[2\pi(\nu_1 - \nu_{s1})t] + E_0 \ \cos[2\pi(\nu_2 + \nu_{s2})t]\} \qquad (4)$$

with $\nu_{s1} = v/\lambda_1$ and $\nu_{s2} = v/\lambda_2$, where v is the velocity to be measured and λ_1 and λ_2 denote the wavelength of the two light waves.

The signal component at the difference frequency is

$$I_D = E_{0-} \cos[2\pi(\nu_1 - \nu_2 - 2\nu_s)t] \quad \text{with} \quad \nu_s \approx \nu_{s1} \approx \nu_{s2}. \qquad (5)$$

Conventional LDA systems use Bragg cells for shifting the original light frequency. A miniaturized and very elegant method for frequency shifting is the sawtooth-shaped phase modulation. This is realized by an electrooptical phase modulator, based on lithiumniobate (LiNbO$_3$) technology, as shown in Figure 4. The light of a laser diode is split in the Io device into two waveguides. At both sides of the outgoing waveguides, planar electrodes are positioned. The refractive index of the material is modulated by the electrical fields within the waveguide region, causing the desired phase shift. The device is also a very efficient polarizer. The polarization of light is due to the fact that the waveguides guide only light of one polarization.

A measurement system as shown in Figure 5 uses a modulation frequency of 20 MHz. The required voltage is 3 V. The device is fiber coupled with polarization, maintaining monomode fibers at the input and at the two output ports. The measurement range is -40 m/s to $+40$ m/s. with an accuracy of 0.1%. By integration, a displacement measurement is performed. The achievable length resolution is 1 mm [4, 5].

4 Rotation Rate Sensing

The inertial detection of rotation rates is based on the so-called Sagnac effect. Due to this phenomenon, a light wave guided through a closed loop suffers a

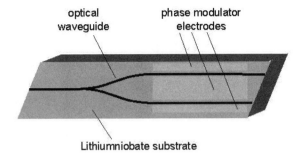

optical
waveguide

phase modulator
electrodes

Lithiumniobate substrate

Fig. 4. Integrated optical beam splitter and phase modulator.

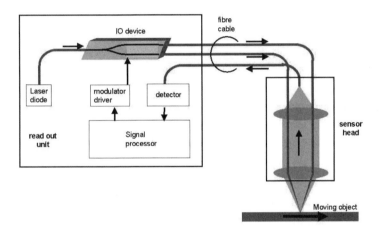

Fig. 5. LDA system with IO phase modulator and miniaturized sensor head.

phase shift of

$$\Delta\varphi = 8\pi N[A/(c\lambda)]\Omega, \qquad (6)$$

where Ω is the rotation rate in arc per second, N is the number of loops, and A is the area within the light path [6, 7].

Standard sensors, based on this effect, are laser gyroscopes, which are used, for example, for flight navigation. Passive devices include fiber gyroscopes, as shown in Figure 6. The light from a laser diode or superluminescent diode is coupled into both ends of a fiber coil, using the integrated optical Y-branch [8]. After having traveled through the fiber, the waves are recombined in the Y-branch and interfere. If the whole arrangement is rotating, a phase shift is observed due to the Sagnac effect. The same IO device device is used as for the LDA sensor. It combines the functions of beam splitting, polarization,

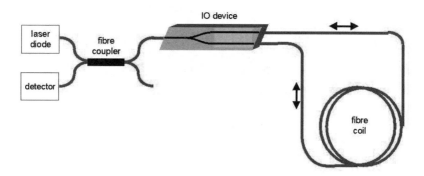

Fig. 6. Passive fiber gyroscope.

and phase modulation. A sinosodial phase modulation is performed in order to yield an output signal, which is proportional to the rotation rate. By filtering the detector signal at the modulation frequency, for small signals, the amplitude of the resulting signal is proportional to the phase difference of the clockwise and counter-clockwise propagating waves. Alternatively, digital signal detection schemes are used. The resolution of fiber gyroscopes ranges from 0.005°/h down to 0.001°/h.

5 Microspectrometer

Small and cost-efficient spectrometers are needed for color measurement, identification purposes, and diagnostic applications. A diode array spectrometer, based on a planar waveguide structure, was realized for the visible, ultraviolet (UV), and the near-infrared (NIR) spectral range (Fig. 7) [9]. The microspectrometer is based on a hollow-cavity waveguide design without any moving parts. It is attached to a photodiode detector array. The light is coupled into the spectrometer through a 300/330 μm (core/cladding diameter) silica fiber and the entrance slit. The light is guided by total reflection inside the spectrometer cavity. The spectrometer itself is a micromolded monolithical device that includes the entrance slit, a focusing flat field echelette grating, and the camera mirror. These elements are arranged in the Rowland design. The microstructures are replicated with optical surface quality using the LIGA technology (Fig. 7) [10]. The monolithic Rowland design guarantees good mechanical, thermal, and optical stability. There is virtually no thermal drift of the wavelength calibration due to the fixed angular relationship between the optical components. State-of-the-art Si detector arrays are used for the UV/visible spectrometer, whereas the NIR microspectrometer incorporates an array made by InGaAs detector technology.

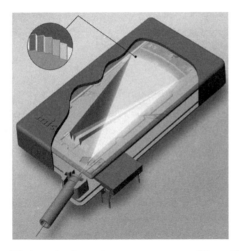

Fig. 7. Miniaturized grating spectrometer. (From Ref. [9])

The UV/visible spectrometer is designed to operate over the UV to the NIR spectral range. In the first diffraction order, it can be used from 320 to 920 nm. In the second diffraction order, the range from 220 to 350 nm is accessible. The spectral resolution is >10 nm at a pixel dispersion of 3.5 nm/pixel. The dynamic range is up to 10,000 : 1 (at 80% full scale). Another spectrometer is designed for the NIR region, covering the spectral range from 1100 nm to 1700 nm. Here, the spectral resolution is >16 nm at a pixel dispersion of 10.5 nm/pixel. There is no requirement for cooling the detector.

The microspectrometers are compact hand-held, battery-powered devices for applications like transcutaneous measurement of blood parameters in medical diagnostics, analysis of specific fluorescence spectra for product identification, colorimetry for quality control of prints, material and light sources, analysis of agricultural and nutrition products, and transmission and reflection measurement of coatings and filters.

6 Fiber-Tip Sensors

Extremely small sensor heads for chemical sensors are realized by coating the tips of optical fibers with chemically sensitive materials. Analogous to the electrodes of electrochemical sensors, these sensors are called "optodes" or "optrodes." The sensitive films usually consist of a membrane material, usually a polymer and an indicator. The indicators, for example, change their transmission when the species to be detected diffuse into the membrane material. Alternatively, fluorescent dyes are used. The change of fluorescence intensity or fluorescence decay time are read out optically [11].

The measurement of oxygen concentration can be done with fluorescing dyes, based on the effect of fluorescence quenching. For many dyes, fluores-

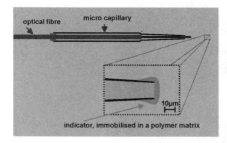 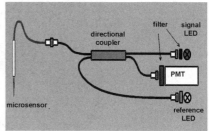

Fig. 8. Oxygen optode, basing on fluorescence decay time. (From Ref. [13])

cence intensity and decay time depend on oxygen concentration according to Stern–Volmer equation:

$$I_0/I = \tau_0/\tau = 1 + k[O_2], \tag{7}$$

where I_0 and I are fluorescence intensity in the presence and in absence of oxygen; τ_0 and τ are luminescence decay time in presence and in absence of oxygen. k and $[O_2]$ are the Stern–Volmer constant and the oxygen concentration, respectively. The fluorescence of several indicator dyes is quenched in the presence of oxygen. Indicators used are, for example, polycyclic aromatic hydrocarbons, transition metal complexes of Ru(II), Os(II), and Rh(II), and phosphorescent porphyrins containing Pt(II) or Pd(II) as the central atom.

The optical arrangement of an oxygen optode is shown schematically in Figure 8; the fluorescence of a coated fiber tip is shown in Figure 9 [12,13].

Fluorescence intensity measurement is strongly affected by all influences, leading to intensity changes within the optical arrangement, including fiber bending losses, photobleaching of the dyes, variations of source intensity, and coupling efficiency. All of these problems are overcome by using the decay measurement, which does not depend on absolute intensity values.

The small dimensions of the fiber tips provide a spatially well-defined measurement. All of these sensors also work as remote sensors with sensitive films deposited on any surface. Excitation and collection of fluorescence light is performed by a fiber end, placed at a distance from the film.

7 Evanescent Wave Sensors

The guided light of a dielectric waveguide is not totally confined to the waveguide. Parts of the waveguide mode are guided outside the surrounding media [14]. This is the so-called evanescent wave, which decays exponentially in the substrate and the superstrate of the waveguide (Fig. 10), allowing the detection of chemical reactions or concentration changes in the vicinity of the

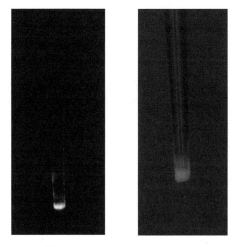

Fig. 9. Fluorescence of a fiber tip, coated with a fluorescing film, left: exitation light, right: fluorescence light. (From Ref. [13])

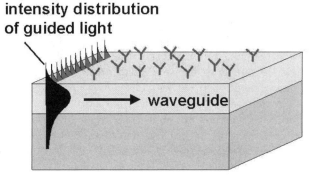

Fig. 10. Intensity distribution of light, guided in a planar waveguide.

waveguide. The evanescent wave method is well suited for the detection of surface-bound chemical reactions or of processes within thin films or small flow cells. The detection volume is very well defined by the extension of the evanescent wave. Only a region of 25 nm to 1 μm, depending on the waveguide and the measuring wavelength, is observed. Background signals from the surrounding volume is efficiently suppressed. Evanescent wave sensing can be extremely sensitive, as the signal is integrated over the surface along the waveguide.

Basically, three types of evanescent field sensor exist: The waveguide photometer detects the absorption of light by a substance on the waveguide surface. With the waveguide fluorometer, fluorescence of a molecule to be detected is excited very efficiently by the evanescent wave. The emitted light can either be collected conventionally by methods of free-space optics or through the waveguide. The third detection principle is the measurement of phase changes, caused by adsorption or binding of molecules on the surface. This can be understood as a change of effective refractive index of the guided wave, caused by the formation of an adlayer with a refractive index, higher than that of the ambient region. The effective refractive index $n_{eff} = c/v_{ph}$ is – analogous to the definition of the refractive index of a homogeneous medium – the ratio of vacuum light velocity c and the mode's phase velocity v_{ph} in the direction of the waveguide.

An example of a waveguide photometer is an infrared sensor for the detection of hydrocarbons in water [15]. The strong absorption lines of organic molecules in the mid-infrared (MIR) between $8\,\mu m$ and $12\,\mu m$ wavelength are used for a specific detection of different species. The sensor is basically a polymer-coated infrared-transparent fiber, made of silver halide. The polymer coating has two functions: The water is kept out of the region of the fiber's evanescent field, as it would cause high additional loss due to the high absorption of water in the MIR. Second, the hydrocarbons not only penetrate the polymer but even enrich the material, increasing the sensor's sensitivity. Because of the integration of light absorption over the fiber length, sensitivity is further increased, as mentioned earlier.

For measurements in bore holes, a compact unit was built, consisting of a light source with fiber coupling optics, the coated fiber, and a spectrometer unit. The spectrum is taken by a small grating spectrometer, covering the spectral range, mentioned earlier. The detection limit, determined, for example, for tetrachloroethylene, is below $1\,ppm$.

The fluorescence excitation through the waveguide is used for the detection of biochemical reactions like DNA - DNA interactions or immune reactions [16]. One of the chemical species is labeled with fluorochrome, which is excited by the evanescent wave. As shown in Figure 11, the waveguide surface can be structured with an array of different recognition molecules (e.g., different oligonucletide sequences). The techniques for the fabrication of large arrays of covalently bound probe molecules are well known from the biochip technology. The fluorescence light is collected by a lens system. With a charge-coupled device (CCD) camera and image recognition software, the intensity is determined of each dot of the array. With cooled CCD cameras, even single molecules can be detected.

The evanescent field detection allows the detection of reaction kinetics (i.e., the time dependence of the binding event), as the fluorescence background from the solved fluorochromes is efficiently suppressed.

The biochemical assay is significantly simplified if the detector can detect the reaction without labels. Labeling of molecules and washing steps become unnecessary. One method of label-free detection is the surface plasmon resonance principle (i.e., the analysis of resonant oscillations of electrons

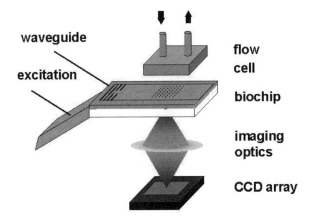

Fig. 11. Waveguide-based fluorescence biochip reader.

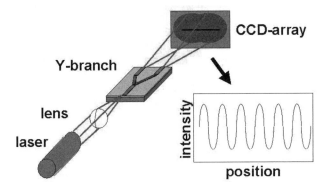

Fig. 12. Interferometric label-free detector.

in a thin metal film) [17]. Basing on dielectric waveguides, label-free detection is possible by monitoring the phase velocity of guided light. The changes of light phase can be directly converted into changes of mass coverage of bound molecules. Two methods for monitoring phase velocity have been proposed: the grating coupler method [18] and interferometric detection [19]. A Young interferometer setup is shown in Figure 12. After splitting the incoming light into two branches – the measuring and the reference branch – the waves couple out and overlap due to the divergent output of the stripe waveguide, forming a pattern of interference fringes. This signature is moving laterally when the phase difference at the output of the two arms varies, indicating a binding event.

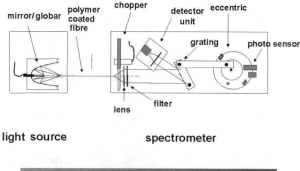

Fig. 13. Miniaturized optical spectrometer. Top: schematic representation of the light path. Bottom: Realized waveguide-optical sensor.

8 Summary

Detection principles of microoptic sensors were discussed. This is not a comprehensive survey, but the most important sensing principles were given with some examples. The sensors or sensor systems, discussed here, are mostly not micrometer-scaled, but their function is essentially determined by the specific properties of a microoptic component.

The following detection principles have been realized using microoptic components: interferometric distance sensing and laser Doppler anemometry are realised in a miniaturized way. Due to their small size, they may be used also in sites which are inaccessible for larger systems. Also, fiber optic rotation sensors are based on an interferometric detection principle. Miniaturized spectrometers provide very small and inexpensive analysis of colours and detection of certain substances in an a sample (Fig. 13). Fiber-optic sensors are used for the detection of chemical species and humidity. Evanescent field sensing provides very sensitive fluorescence detection of biochemical reactions at the surface of a waveguide. Furthermore, this principle allows the detection without labels like fluorescence dyes or radioactive markers.

References

1. Marguerre, H., Optical phase sensitive detection, in *Sensors, Vol 6: Optical Sensors*, edited by Wagner, E., Dändliker, R., and Spenner, K., VCH, Weinheim (1992).

2. Erbeia, C, Valette, S., Jadot, J.P., Gidon, P., and Reard, S., *Integrated Optics Displacement Sensor Connected with Optical Fibres*, Springer Proceedings in Physics, Vol. 44, edited by Arditty, H.J., Dakin, J.P., and Kersten, R.Th., Springer-Verlag Berlin (1989).

3. Ulbers, G., A sensor for dimensional metrology with an interferometer, in *Integrated Optics Technology*, Springer Proceedings in Physics, Vol. 44, edited by Arditty, H.J., Dakin, J.P., and Kersten, R.Th., Springer-Verlag Berlin (1989).

4. Bauer, J., Dammann, E., Fritsch, E., and Rasch, A., Integrated optical three beam interferometer for displacement measurements, Int. J. Optoelectron. 8(5/6), 547–553 (1993).

5. Pradel, T., Rasch, A., Rothardt, M., Tropea, C., Varro, W., and Weber, H., Use of integrated optical devices in laser doppler anemometry, SENSOR 93, Kongreßband III, B 9.1, pp. 179–186 (1993).

6. Böhm, K. and Rodloff, R., Optical rotation sensors, in *Sensors, Vol. 6: Optical Sensors*, edited by Wagner, E., Dändliker, R., and Spenner, K., VCH, Weinheim (1992).

7. Ezekiel, S. and Arditty, H.J., *Fibre-Optic Rotation Sensors and Related Technologies*, Springer-Verlag, Berlin (1982).

8. Rasch, A. and Handrich, E., Applications of lithium niobate and potassium titanyl phosphate integrated optic devices , SPIE Proc. Integrated Optics Devices III, Vol. 3620, Bellingham, WA, pp. 152–160 (1999).

9. Decker, A. and Steag Microparts GmbH, private communication.

10. Mohr, J., Last, A., and Wallrabe, U., Modular fabrication concept for LIGA-based microoptics, this book

11. Wolfbeis, O.S. (ed.), *Fibre Optic Chemical Sensors and Biosensors*, CRC, Boca Raton, FL, Vols. 1 and 2, (1991).

12. Klimant, I., Kühl, M., Glud, R.N., and Holst, G., Optical measurements of oxygen and temperature in microscale: strategies and biological applications, Sensors Actuators B 38-39, 29–37 (1997).

13. Wünschiers, R., Borzner, S., and Stangelmayer, A., Optische Mikrosensoren zur Sauerstoffmessung, in *Chemie in Labor und Biotechnik* 50 209–214 (1999).

14. Hunsperger, R.G., *Integrated optics: Theory and Technology*, Springer-Verlag, Berlin (1982).

15. Hahn, P., Tacke, M., Jakusch, M., Mizaikoff, B., Spector, O., and Ketzir, A., Detection of hydrocarbons in water by MIR evanescent wave spectroscopy with flattened silver halide fibres, Appl. Spectros. 55, 39–43 (2001).

16. Duveneck, G.L., Pawlack, M., Neuschäfer, D., Bär, E., Budach, W., Pieles, U., and Ehrat, M., Novel bioaffinity sensors for trace analysis based on lunescence excitation by planar waveguides, Sensors Actuators B 38–39, 88–95 (1997).

17. Liedberg, B., Nylander, C., and Lundström, I., Surface plasmon resonance for gas detection and biosensing, Sensors Actuators 4, 299–304 (1983).

18. Nellen, Ph., Tiefenthaler, K., and Lukosz, W., Integrated optical input grating couplers as input couplers as biochemical sensors, Sensors Actuators 15, 285–295 (1988).

19. Brandenburg, A., Krauter, R., Künzel, Ch., Stefan, M., and Schulte, H., Interferometric detection of surface-bound bioreactions, Appl. Opt. 39, 6396–6405 (2000).

Fiber-Optic Gyros and MEMS Accelerometers

Andreas Rasch, Eberhard Handrich, Günter Spahlinger, Martin Hafen, Sven Voigt, and Michael Weingärtner

LITEF GmbH Freiburg, P. O. Box 774, 79007 Freiburg, Germany
rasch.andreas@litef.de

Fiber-optic gyros (FOGs) and micro-electro-mechanical-systems (MEMS) accelerometers are used today in inertial strapdown systems for medium accuracy and expanding into high-performance strapdown navigation systems in competition with ring laser gyros (RLGs), whereas from the low-accuracy side, MEMS gyros are used for expanding to the medium accuracy ranges. The FOG principle is based on constant light velocity. This results in a phase difference of lights which are propagating through a fiber coil in clockwise (cw) or counterclockwise (ccw) directions if a rate is applied. The phase difference is proportional to the rate. The FOG technology has been developed from an open-loop design – still used in some market niches – to closed-loop design with high bandwidth and random phase modulation technique. The first generation of FOG systems uses one light source split by a 3×3 coupler to three fiber coils. More than 15.000 FOGs for such triad systems have been produced and delivered. Typical applications are Attitude and Heading Reference Systems or Land Navigators, which are described. The second generation of FOG systems uses single-axis FOGs with internal processors. A large quantity of these fiber-optic rate sensors (µ-FORS) can be easily calibrated separately and later assembled to modular systems. The features of the µ-FORS family for bias values from 6°/h down to 0.05°/h are given. The different bias values are realized by adapting the fiber length on the coil. The other optical parts and the electronics are unchanged. One main feature for the common electronics is the tracking of the modulation frequency to the actual fiber length.

The MEMS accelerometers are still mechanical sensors built by the micro-machining technique. The technology, the mechanical sensor, and the electronics are described on the example of the B-290 Triad, which is a typical MEMS accelerometer product. Test data for bias repeatability and stability and scale factor accuracy before and after temperature compensation are presented.

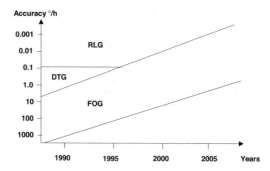

Fig. 1. Evolution of gyro technologies.

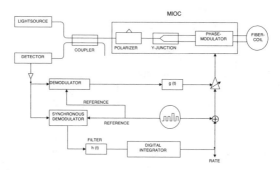

Fig. 2. Block diagram of the closed-loop FOG.

1 Basic Fiber-Optic Gyro Technology

If we look at FOGs, there exist the open- and the closed-loop designs. The major advantage of the closed-loop FOG is the high linearity of the scale factor and its insensitiveness against environment, especially against vibration.

The evolution of the most important gyro technologies in terms of accuracies over the years is shown in Figure 1.

Today, the high-accuracy end is still occupied by RLGs. The moderate accuracy range from 0.01°/h to 30°/h is mainly covered by FOGs, and MEMS gyros are rising up from the low end accuracy.

LITEF produces a closed-loop design only. The block diagram of the closed-loop FOG is shown in Figure 2. The heart of the FOG is the multi-function integrated optic chip (MIOC). The MIOC realizes the polarizer, the main coupler, and the modulator in one chip, which is shown in Figure 3. The MIOC production is done on $LiNbO_3$ wafers by forming proton-exchanged waveguides and sputtered electrodes. Thirty-two MIOCs are diced out of one wafer, which is shown in Figure 4.

The advantage of this technology is the high extinction ratio of the polarizing waveguides formed by proton exchange [1]. The drawback of this technology is the high up-front investment for an independent in-house pro-

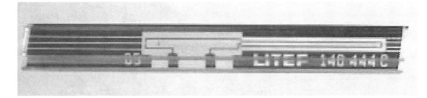

Fig. 3. Multifunction integrated optic chip (MIOC).

Fig. 4. Lithium niobate technology steps: 3-in. wafer with 32 MIOCs, subwafers, and single chips.

duction for the MIOC and the challenge of the new technology, which requires clean rooms and equipment for lithography, proton-exchange baths, annealing ovens, sputtering equipment, wafer dicing tools, and chip polishing tools.

Between 1994 and 2002, LITEF produced more than 30.000 MIOCs with high yield for its FOG products.

2 FOG-Triad Systems

At the beginning of the 1990s, the superluminescent light-emitting diode (SLD) light source package was a question of cost for fiber-optic production. Therefore, the natural decision was to use one SLD light source and distribute the light power to three IO chips and fiber coils. Such a triad structure is shown in Figure 5; a typical sensor block assembly is shown in Figure 6 – this may be used in the Attitude and Heading Reference System (AHRS) for commercial applications.

To date, LITEF has developed five different triad configuration systems:

- LITEF Commercial Reference-92 (LCR-92), an Attitude Heading Reference System with bubbles as the level sensor for commercial airborne applications
- LITEF Commercial Reference-93 (LCR-93), an Attitude and Heading Reference System with integrated silicon accelerometers for commercial airborne applications

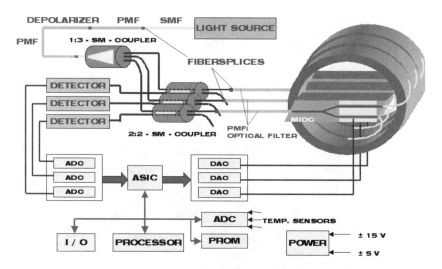

Fig. 5. Fiber-optic gyroscope – triade structure.

Fig. 6. Sensor block micro Attitude and Heading Referencs System.

- LITEF Transport Reference-97 (LTR-97), an Attitude and Heading Reference System with bubbles as the level sensor for airline and transport application
- LITEF Land Navigator-GX (LLN-GX), navigator, which integrates the information of the FOGs, of the bubbles, of the odometer of the vehicle, and of a GPS receiver to optimal navigation data
- LITEF Land Navigator-G1 (LLN-G1), a navigator similar to LLN-GX but with high accuracy FOGs, with accelerometers instead of bubbles and integrated with self-alignment features to the north

The accuracy span from 3°/h to 0.08°/h of those systems is achieved with different coil designs and fiber lengths. The electronics are almost identical. All of these systems have been in production for several years and the quantities produced have increased; for example, between January 2000 and July 2002, more than 3000 Triad Fiber-Optic Systems were produced. Therefore, LITEF has gained much experience in the field of FOG production yield. Where production yield is concerned, the most critical production test is calibration over the temperature of the system, which is done on a turntable with a climate chamber. Such tests are fully automatic and steered by computers, and four systems can be calibrated simultaneously; however, the test time is between 10 and 36 h and the test equipment is expensive. Today, the production yield in calibration of almost all Fiber-Optic Triad Systems is over 90 %. In addition to the well-known measurement for optical reciprocity and random phase modulation with its auxiliary loops, the main problems to be overcome for that yield were as follows:

- Electronic noise at the IO modulator that can create bias errors that are difficult to describe with a model. Filtering is very limited because of the required high bandwidth of 100 MHz for the modulation.
- Effects on bias by the so-called "bunny ears" created by electronic transients in conjunction with the interferometer transfer function. Nonlinearities in the detector and amplifier channel can also create bias.
- Wavelength-selective optical losses in fiber and couplers that create scale factor problems over temperature.

To always be able to calibrate three FOG axes simultaneously, a high-performance margin is required for the FOGs to achieve the 90 % yield in production.

3 Modular Fiber-Optic System Design

Modular system design is a general design required for an easy assembly of a system in production. This is also realized in Fiber-Optic Triad Systems. However, a higher level of modularity means that each component of such a system should be testable in its function with high failure elimination, not only for an easy assembly but also for a successful test (e.g., calibration and acceptance test). This is difficult to achieve for Fiber-Optic Gyro Triad Systems, because the total function (e.g., bias or scale factor accuracy) can only be tested with about 50% yield on the component level. In calibration, all parameters of all three axes then have to be tested simultaneously, which creates high requirements for the material and functionality of components and their integration. However, such a modular system design can easily be realized with single-axis FOGs with internal electronics and a processor which compensates for bias and scale factor over temperature. Each gyro can be calibrated individually

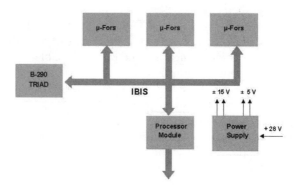

Fig. 7. Modular FOG inertial measurement unit (IMU) approach.

over temperature and later assembled within systems with orthogonal or redundant skewed axes. A block diagram of such a modular system is shown in Figure 7.

A digital synchronous bus IBIS (Intelligent Bus for Inertial Sensors) links the single-axis micro-fiber-optic rate sensors (μ-FORS) and the triad including silicon accelerometers (B-290 Triad) with the processor module. The data rate can be programmed between 5 Hz and 8 kHz. A power supply can be added if the required voltages (±15 V and ±5 V) are not delivered.

4 Single-Axis μ-FORS Family

The μ-FORS was developed in 1995 [2] and more than 6000 μ-FORS have been produced and delivered to date. μ-FORS is a single-axis fiber-optic rate sensor with the necessary optics and electronics in a small housing of $76 \times 55 \times 20$ mm^3. It requires ±5 V and 2 W and delivers the rate data in digital format via the IBIS bus. Bias and scale factor are compensated over temperature internally by a processor within the digital application specified integrated circuit (ASIC). Many features can be programmed (e.g., the rate range and the output data rate) [3]. The μ-FORS uses a SLD without Peltier cooler, a binary digital MIOC with integrated digital-to-analog converter (DAC) function, a low-cost detector, and a flash ADC. The main control loop, all auxiliary loops,

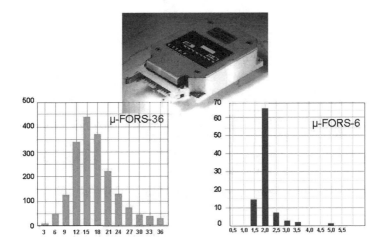

Fig. 8. Bias statistics of μ-FORS-36 and μ-FORS 6.

and the reduced instruction set computer (RISK) processor are to be incorporated within a digital ASIC with 1.4 Mio transistors. The main bulk of the production quantities are dedicated to μ-FORS-36 and μ-FORS-6, the bias performance which is shown in Figure 8.

In the meantime, LITEF has developed a μ-FORS family with the following features:

• Compact single-axis rate sensors
• Temperature-compensated digital output
• Programmable rate range, with higher resolution and data rate
• Fiber coil length from 50 m to 500 m
• Bias residual errors from 6°/h to 0.03°/h
• Scale factor error from 3000 ppm to 100 ppm
• Typical size: 76 mm×55 mm×20 mm

The major steps for the improvement of bias and scale factor to these limits have been a new digital MIOC with 12-bit electrodes and a new digital ASIC with 4-Mio transistors, which includes the following improvements [4]:

• An improved hardware scale factor control
• An optical coil fiber length measurement
• Mdulation frequency tracking to the actual fiber length
• A bit-weighting compensation algorithm for the digital MIOC
• Subsequent noise and resolution reduction
• Improved data path for main control loop
• ARM 7 RISK processor with RAM/ROM

Fig. 9. Optical components of μ-FORS with a 500-m fiber coil.

Fig. 10. MIOC and electronics board of the μ-FORS.

With the new feature – the tracking of the modulation frequency to the actual fiber length – the bias effects, created by synchronous noise, vanish is a significant improvement.

The smallest μ-FORS family member is the μ-FORS-36m. The main challenge was the development of an analog ASIC together with a multichip module for the detector and an optimized packaging of optics and electronics.

The most accurate μ-FORS has a 500-m fiber coil and a Peltier stabilized SLD. This μ-FORS is shown in Figures 9 and 10.

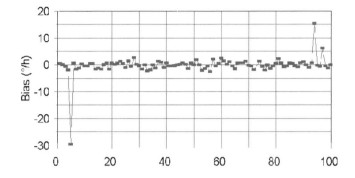

Fig. 11. Measured bias values of μ-FORS 36 m at room temperature.

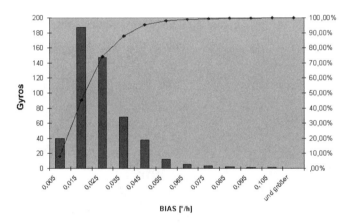

Fig. 12. Bias statistics of a 500-m FOG.

Figure 11 showes the measured bias values of 100 sensors μ-FORS-36m. It can be seen that all values except three axes are within $\pm 3°/h$.

The best performance of 500 m FOGs in production is shown in the bias statistics of Figure 12; 90 % of the produced gyros are below 0.04°/h bias and the typical bias (highest peak)is 0.015°/h.

5 MEMS Accelerometer

Most of the MEMS accelerometers principles use elastically supported pendulums produced by bulk micromachining. A typical example of that type is LITEF's B-290 accelerometer whose principle is shown in Figure 13.

Upon acceleration, the position of the pendulum and the gaps between the cover wafers are changed. The gaps are used as a capacity bridge for the

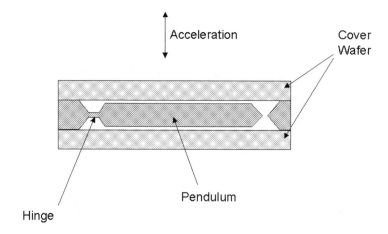

Fig. 13. Principle of the silicon accelerometer B-290.

Fig. 14. Silicon accelerometer chip for B-290.

pick-off and as the restoring torquer by means of electrostatic forces. The total silicon chip is shown in Figure 14. The chip is built out of five silicon wafers by silicon direct bonding: Two shield wafers are added to avoid stray capacities.

The opened chip allows the view of the pendulum with the hinges and the electrode with the shielding frame on the other side in Figure 15. The production process of the chip is actually a batch process on 5 wafers for 140 chips which are bonded together at the end of the process and then cut into 140 accelerometer chips.

The sensor electronics switch the voltage at the capacitor bridge and sense the differences in the capacitor bridge by a charge amplifier and an ADC. A

Fig. 15. Open accelerometer chip for B-290.

Fig. 16. Open B-290 accelerometer triad.

signal processor performs the linearization and steers the restoring by electrostatic forces. The digital acceleration output of the processor is compensated in bias and scale factor over temperature. The electronics for a sensor are integrated within two small hybrids; three sensors and three electronic sets are built into a triad, as shown in Figure 16.

In a cost-reduction program, a mixed signal ASIC was developed for the charge amplifier, different voltage controls, and the ADC used for each sensor. Only one signal processor is used for three sensors, which dropped cost and power consumption of the B-290 triad. The new B-290 triad is shown in Figure 17.

Each B-290 triad is calibrated over temperature and scale factor and bias is compensated inside by the signal processor. Typical scale factor and bias repeatability over temperature are shown in Figures 18 and 19.

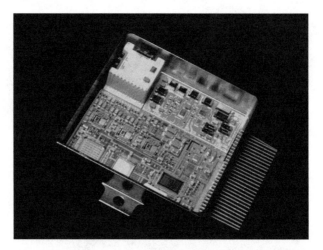

Fig. 17. New design of the B-290 accelerometer triad.

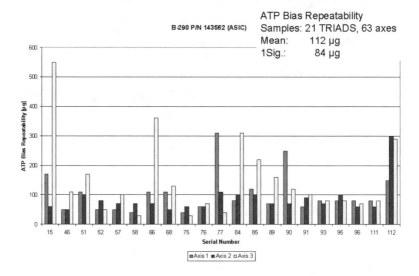

Fig. 18. Bias repeatability over temperature.

6 Inertial Measurement Unit

Three μ-FORS and one B-290 triad are the inertial sensors for an orthogonal inertial measurement unit (IMU); only the processor and perhaps a separate power supply has to be added according to the block diagram of Figure 7. Due to separation into single-axis FOGs, it is easy to adapt the housing for different geometric requirements. Normally, this packaging is carried out by the manufacturer, but this may also be done by the customer. As an example,

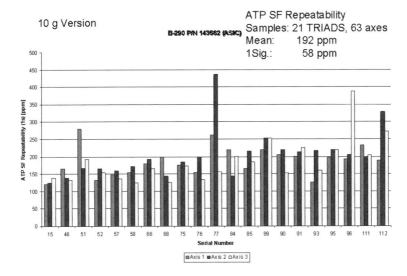

Fig. 19. Scale factor repeatability over temperature.

Fig. 20. Example of a modular measurement unit.

a packaged IMU is shown in Figure 20, with integrated IMU processor and power supply. The size is $13 \times 11 \times 7.5 \, \text{cm}^3$ and the weight is $1.1 \, \text{kg}$.

7 Conclusion

The chapter presents two generations of FOG systems by LITEF. The first FOG system generation consists of one light source split by a 3×3 coupler to three fiber coils. The second generation of FOG systems uses single-axis FOGs with an internal processor. These fiber-optic rate sensors can easily be calibrated separately in high quantity. The μ-FORS family is characterized by bias errors ranging from 6°/h to 0.05°/h. The different bias values are realized by adapting the fiber length on the coil. The other optical parts and the electronis are unchanged.

The mechanical sensor, including electronics, and the underlying technology are described by the example of the B-290 triad, a typical MEMS accelerometer product. Test data for bias repeatability and stability and scale factor accuracy before and after temperature compensation are presented.

Three μ-FORS and one B-290 triad are the inertial sensors for an orthogonal IMU, only the processor and perhaps a seperate power supply has to be added.

References

1. Rasch, A. and Handrich, E., Applications of lithium niobate and potassium titanyl phosphate integrated optic devices, SPIE Proc. 3620, 152–160 (1999).
2. Kemmler, M. et al., Design of a commercial small-volume fiber-optic gyro, SPIE Proc. 2837, 92–97 (1996).
3. Spahlinger, G. et al., Error compensation via signal correlation in high precision closed loop fiber-optic gyros, Fiber-Optic Gyros: 20th Anniversary Conference, Session 3, Denver, 1996.
4. Kunz, J. et al., Design of an ASIC for a commercial small-volume fiber-optic gyro, Fiber-Optic Gyros: 20th Anniversary Conference, Session 3, Denver, 1996.

Microoptical Applications in Spectrally Encoded Fiber Sensor Systems

Hartmut Bartelt

Institut für Physikalische Hochtechnologie, Jena, Albert-Einstein-Strasse 9, 07745 Jena, Germany
hartmut.bartelt@ipht-jena.de

1 Introduction

Miniaturization is a general trend in many technical applications. In addition to the dramatic development in microelectronics, optoelectronic and photonic systems have largely participated in this development. Fiber sensor systems are highly illustrative examples not only of the use of microoptical components but also of the application of microsystems technology in general, including optoelectronic and micromechanical components [1–3]. In this chapter, we will discuss the field of spectrally encoded fiber sensor systems in this context.

The development of high-quality and low-cost optical fibers for telecommunication applications has also inspired the rapidly growing field of fiber-optical sensors. For such fiber sensors, a distinction can be made between intrinsic concepts, where the fiber itself serves as a sensitive element, and extrinsic concepts, where the fiber serves only as an optical communication link. Generally, all optical parameters defining the state of light, such as amplitude, phase, wavelength, or polarization, can be used for measuring an external physical quantity via an optical fiber sensor.

The most direct way of designing a fiber sensor element is based on the variation of the optical amplitude (intensity-modulated sensor), since many parameters are known to influence the transmission properties of optical fibers. However, such a concept usually suffers from strong cross-sensitivity; moreover, it is dependent on several component properties (light source and detector variations, coupling, transmission attenuation, etc.). The usefulness of such a concept is limited therefore. A sensor encoding concept based on spectral modulation instead of amplitude or intensity modulation largely reduces such problems. All attenuation effects that do not change the relative intensity of different wavelength channels will not degrade the accuracy of the sensor. Spectrally encoded fiber sensor systems are, therefore, especially interesting for applications where a reasonable degree of accuracy is required. In addition, such sensors make it possible to exploit other general advantages of fiber sensor systems as well, such as small size and weight, immunity against electromagnetic radiation, electrically insulating properties (e.g., for high-voltage applications), or applicability under adverse environmental conditions.

In the following, a variety of different spectral effects and their implementations as a fiber microoptical sensor are presented, including spectral scattering, spectral attenuation or absorption, and spectral reflection effects.

2 Spectral Encoding Concepts

2.1 Raman and Brillouin Scattering

Raman and Brillouin scattering are nonlinear effects related to inelastic scattering of photons involving vibrational states or optical phonons (Raman scattering) or acoustical waves or acoustical scattering (Brillouin scattering). In both cases, the scattered signal is spectrally shifted compared to the original signal. Although silica is not intrinsically a highly nonlinear material, considerable nonlinear effects may be observed in such optical fibers due to the possible high energy density in a fiber core. The shift in wavelength depends on parameters such as temperature and enables the design of fiber sensors. To this end, light is coupled into the fiber and the scattered light is spectrally analyzed. Typically, pulsed light is used, which allows local resolution at position z based on the measurement of the traveling time dt of the optical pulse (optical time-domain reflectometry, OTDR):

$$z = 2c\,dt, \tag{1}$$

where c is the speed of light. With a time resolution in the nanosecond rage, spatial resolution in the meter range is generally possible.

Raman scattering is relatively weak and gives two scattering peaks (Stokes and anti-Stokes at wavelengths λ_s and λ_a, respectively) with a shifted frequency of about 14 THz. The relative intensity of the two shifted Raman scattering lines depends on the temperature according to [1]

$$R(T) = (\lambda_s/\lambda_a)^4 \exp(-h\nu/kT), \tag{2}$$

where T is the absolute temperature, h is Plank's constant, k is Boltzmann's constant, and ν is the frequency of incident light. This relation allows a temperature measurement by comparing both Raman components and is implemented in commercial backscattering measurement systems.

The scattered Brillouin intensity is relatively high compared to Raman scattering, but the typically observed frequency shift ν_B corresponds to only about 12 GHz, depending on the optical material. Due to the acoustical properties relevant for the frequency shift of the scattered light, this frequency shift ν_B will vary with temperature T and strain ϵ of the optical fiber and can be approximately described as

temperature

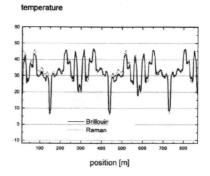

position [m]

Fig. 2. Comparison of Raman and Brillouin temperature measurement along an electrical cable.

Fig. 1. Brillouin measurement setup.

$$\nu_B = (2n v_A)/\lambda, \tag{3}$$

where n is the refractive index, v_A is the acoustical wave velocity, and λ is the wavelength.

The sensitivity C of such a sensor is typically determined by the relative change of the frequency shift and is given by

$$C_s = 1/\nu_B (d\nu_B/d\epsilon) \tag{4}$$

(typical value 4.5),

$$C_t = 1/\nu_B (d\nu_B/dT) \tag{5}$$

(typical value 10^{-4} K^{-1}).

A change of the frequency shift involves the refractive index and the acoustical velocity. The variation of the acoustical velocity is the dominant effect compared to the variation of the refractive index. The relation between the physical quantity and the frequency shift is linear to a good approximation and allows the implementation of distributed strain and temperature sensing (Figs. 1 and 2) [4].

2.2 Spectral Absorption and Evanescent Wave Spectroscopy

Spectral absorption properties are valuable parameters for characterizing materials or material properties. Such properties may be measured with fiber optical systems either using a separate, miniaturized measuring cell or employing the evanescent fields directly around a fiber. In the case of evanescent wave spectroscopy, the sensitive volume is situated around the fiber and close to the fiber core. The penetration depth of the evanescent field is typically in the range of an optical wavelength and the exponential decay can be described as [1]

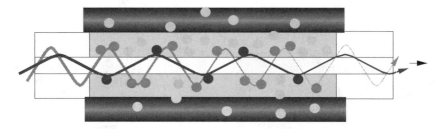

Fig. 3. Polymer-coated fiber for evanescent field sensing.

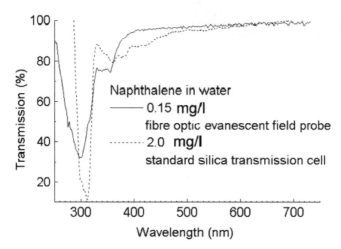

Fig. 4. Naphtalene sensing in water with an evanescent fiber sensor in comparison to a standard transmission cell.

$$d_p = \lambda/[2\pi(n_1 \sin\Theta - n_2)^{1/2}], \qquad (6)$$

where n_1 is the higher-refractive-index medium, n_2 is the lower-refractive-index medium, and Θ is the incidence angle.

The concept of such fiber-optical evanescent wave spectroscopy allows for long interaction lengths and is relatively resistant to dirt layers, since the optical transmission channel (fiber core) is not affected by such processes. The fiber concept is generally applicable to the investigation of gaseous or fluid materials. In order to get access to the evanescent wave, either the fiber coating has to be thinned down to the micrometer range or the coating must allow penetration of the molecules or atoms under investigation. In this case, effects of enrichment in the coating layer can be used to increase sensitivity (Fig. 3). As an example, polymer layers have been successfully applied for the measurement of hydrocarbonates in water. An example for the measurement of naphtalene is shown in Figure 4.

2.3 Spectral Reflectivity

Variable mirror structures in combination with optical fibers can be used for the measurement of a great variety of physical quantities. Fabry–Perot filters are often used as such spectrum-modifying elements. They consist of two mirror layers separated by a distance layer. The spectral characteristics of such an element is influenced by the reflectivities of the two mirror layers and the optical path length difference between them. In the case of equal reflectivity for both mirrors, the transmitted intensity is described by [1]

$$I = I_0/(1 + F \sin{\_} \Phi/2), \tag{7}$$

where $F = 4R/(1 - R)$, R is the reflectivity of a single mirror, and Φ is the optical phase shift in the resonator.

The phase retardance Φ can be changed either by varying the path length difference (distance of mirrors) or by varying the effective index of the material between the mirrors. Depending on the strength of reflection, either a wavelength-dependent intensity characteristic similar to a cosine function or an intensity characteristic with a rather discrete array of maxima is achieved. A measure of sensor sensitivity is the change of intensity dI with phase change $d\Phi$. The maximum sensitivity is achieved for $\Phi_- = 4/3\, F$ and reaches the value of $(-0.3\, I_0 F^{1/2})$ [1].

Some typical examples for parameters to change a Fabry–Perot element are temperature (distance or refractive index change), force or pressure (distance change of a membrane), or humidity (refractive-index change). In Figures 5 and 6, a pressure sensor element is shown, where the resonator is achieved by combination of a mircrostructured silicon membrane and a quartz block. In the case of the humidity sensor (Figs. 7 and 8), a microporous layer system is used as the filter. Since the size of the pores is in the subwavelength range, the optical performance is not influenced by scattering. Condensed water (or eventually other gases/liquids) changes the effective index and, therefore, the optical path difference. Due to the small volumes involved, this process of index change is fast and reversible and allows for measurement also in the very low-humidity range.

Fiber Bragg gratings represent an alternative wavelength-specific mirror structure which is very well suited for wavelength-encoded sensor functions [5,6]. A fiber Bragg grating is implemented as a modulation of the refractive index within the core of an optical fiber, typically by illumination with a periodic pattern at ultraviolet (UV) wavelengths (Fig. 9). Such a modulation pattern within the fiber core may be described as

$$\Delta n(z) = \Delta n_{DC}(z) + \Delta n_{AC}(z) \cos[2\pi z/\Lambda_{FBG} + \Phi(z)], \tag{8}$$

where Λ_{FBG} is the grating period, Δn_{DC} is the bulk refractive-index change, Δn_{AC} is the refractive-index modulation, and Φ is the phase shift in the grating structure.

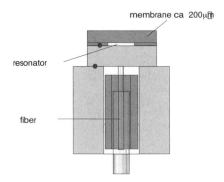

membrane ca 200μm

resonator

fiber

Fig. 5. Setup for a fiber-optical pressure sensor based on a Fabry–Perot filter (schematic).

Fig. 6. Implemented optical module of a fiber-optical pressure sensor.

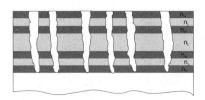

Fig. 7. Principle of a layered porous Fabry–Perot filter for a humidity sensor.

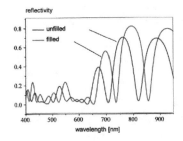

Fig. 8. Humidity measurement with the Fabry–Perot filter of Figure 7 at an optical fiber end.

A large refractive-index modulation is achieved not only by the high exposure dose but also by high photosensitiviy due to doping of the fibers or appropriate preprocessing. Values in the range of 10^{-5} up to 10^{-2} are achieved. Either a phase mask method or an interferometric method is used as a recording scheme. The interferometric method allows a very flexible adjustment of the Bragg period and is applied, for example, in combination with a fiber drawing tower for the realization of grating arrays (Fig. 10).

The reflection peak is typically very narrow (in the range of 1 nm and less) and may reach reflectivities close to 100%. A typical example for such a fiber Bragg grating reflection peak is shown in Figure 11. The design wavelength of the reflection peak depends on the effective index and the grating period:

$$\lambda_D = 2n_{eff}\Lambda_{FBG}, \tag{9}$$

where n_{eff} is the effective refractive index and Λ_{FBG} is the period of the fiber Bragg grating.

The design wavelength may be changed by external parameters and will result in an effective Bragg wavelength λ_B of the reflection peak, which depends

grating period ~1/2000 mm

fiber cladding
fiber core
(5-10μm)
refractive index

Fig. 9. Principle of fiber Bragg gratings.

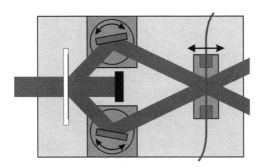

Fig. 10. Talbot interferometer for interferometric recording of fiber Bragg gratings with a phase grating as a beam splitter applied at a fiber-drawing tower.

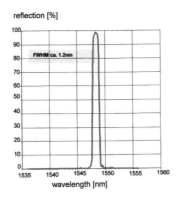

Fig. 11. Reflection characteristic of a fiber Bragg grating.

on the bulk index refraction change Δn_{DC} according to

$$(\lambda_B - \lambda_D)/\lambda_D = \Delta n_{DC}/n_{eff}. \tag{10}$$

The period of the Bragg grating and the refractive index can be changed by physical parameters such as temperature or mechanical strain. Accordingly, the reflection peak will be changed in wavelength. The shift is relatively

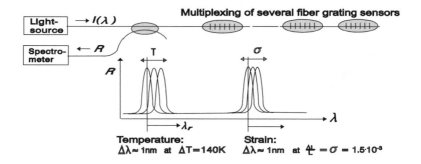

Fig. 12. Fiber Bragg gratings as sensor elements for temperature and strain.

small (in the range of 1 nm or less) but sufficient for sensor measurements (Fig. 12). The concept of such a fiber Bragg grating allows multiplexing of sensors in a grating array, where specific sensor positions can be distinguished by wavelength or delay time in a pulsed system [7].

3 Conclusions

Microoptical components are essential parts in fiber-optical sensors. This includes not only the fiber itself but also its sensor-specific modifications, as well as coupling optics or optical sensor modules. Miniaturization and stability are essential aspects for application. Concepts using wavelength encoding offer attractive system properties. The development of new solid-state laser light sources and compact spectral analyzing systems will further intensify this trend [8].

Acknowledgments

Helpful discussions and contributions by Prof. R. Willsch, Dr. W. Ecke, Dr. G. Schwotzer, Dr. V. Hagemann, Dr. S. Schroeter, C. Chojetzki, H. Lehmann, and T. Widuwilt as well as funding by the Federal Ministry of Education and Research (BMBF) are gratefully acknowledged.

References

1. Dakin, J. and Culshaw, B., (eds.), *Optical Fibre Sensors*, Artech House, Norwood, MA (1988/89), Vols. I–III.
2. Rogers, A.J., Optical-fiber sensors, in *Sensors, Vol. 6, Optical Sensors*, edited by Göpel, W., Hesse, J., and Zemel, J.N., VCH, Weinheim, (1992).

3. Bartelt, H., Spectrally encoded fiber sensor systems, Proc. SPIE 4900, 394–400 (2002).
4. Geinitz, E., Jetschke, S., Röpke, U., Schröter, S., Willsch, R., and Bartelt, H., The influence of pulse amplification on distributed fibre-optic Brillouin sensing and a method to compensate for systematic errors, Meas. Sci. Technol. 10, 112–116 (1999).
5. Kersey, A.D. et al., Fiber grating sensors, J. Lightwave Technol. 15, 1442–1463 (1997).
6. Rao, Y.J., In-fibre Bragg grating sensors, Meas. Sci. Technol. 8, 355–375 (1997).
7. Bartelt, H., Grimm, S., Hagemann, V., Rothhardt, M., Ecke, W., and Willsch, R., From photosensitive fibers to fiber Bragg grating sensor systems, Proc. SPIE 4900, 424–429 (2002).
8. Willsch, R., Application of optical fibre sensors: technical and market trends, Proc. SPIE 4074, 24–31 (2000).

Microoptical Beam Shaping for Supershort-Pulse Lasers

Rüdiger Grunwald[1] and Volker Kebbel[2]

[1] Max-Born-Institute for Nonlinear Optics and Short-Pulse Spectroscopy,
Max-Born-Straße 2a, D-12489 Berlin, Germany
grunwald@mbi-berlin.de
[2] Bremen Institute for Applied Beam Technology, Klagenfurter Straße 2, D-28359
Bremen, Germany

1 Introduction

One of the most exciting ideas in the field of modern optics is the combination of *microoptical components* with *ultrashort laser pulses*. In this way, the specific advantages of both (miniaturization in spatial dimensions and localization in time) can be exploited for novel-type devices.

Ultrashort high-power lasers are of increasing interest for many advanced applications in chemistry [1], nonlinear optics, materials processing, medicine, biology, and information technology because of the new prospects for selective excitation, efficient frequency conversion, access to higher-order nonlinearities, particle manipulation, and the potential of high-speed optical data processing [2]. On a sub-10-fs timescale, a sufficiently accurate control of the beam properties of laser sources is difficult. In the optical or near-infrared part of the spectrum, such pulses can be described by wave packets with a duration even below a single oscillation cycle. The corresponding spectrum is very broad, may span a full optical octave and more for the shortest pulses, and typically contains distinct substructures. In addition, the pulse may have a significant frequency chirp. In contrast to longer pulses, spatial and temporal parameters are strongly coupled ("space-time coupling") and a spatial chirp has to be taken into account [3]. Therefore, any shaping or characterization procedure has to be designed with respect to both a sufficiently high spatial and temporal resolution. Furthermore, the propagation is significantly influenced by spectral dispersion as well as spatially induced group velocity dispersion by amplifier or compressor systems, light guiding systems, and even the air between. Therefore, the optimization of the pulse structure and the development of the necessary diagnostic techniques are closely connected.

A very promising approach to solve these problems is the use of novel types of low-dispersion *thin-film microoptical component* [4–6]. Among the unique features of this kind of microstructures, their ability to realize very small phase gradients is of extraordinary interest for ultrashort-pulse optics. The access to angular resolutions far below 1° combined with sophisticated array geometries and defined local phase functions allows the generation of highly structured light patterns. Thin-film microlenses with *nonspherical shape functions* (axicons [7]) allow one to produce multiple or extended foci [8, 9] and are less sensitive against tilt or axial displacement, like spherical lenses. With thin refractive microaxicons of conical or Gaussian shape, *nondiffracting beams* [10] of Bessel and Bessel-like [11] intensity distribution can be generated by constructive interference of conical partial beams. A conversion of information from the temporal domain into the spatial domain (spatial frequency spectrum) can be realized with a fringe frequency well adapted to the system dimensions (subaperture diameters) and to the application. With arrays of such microaxicons, a ray optical approximation of wave optics can be directly implemented experimentally. Beams of extremely low spatial frequencies are possible (e.g., strongly degenerated Bessel-like beams with only a single maximum but extended depth of focus). Additional degrees of freedom for complex wave shaping can be introduced by spatially graded stacks of different dielectric materials (*multilayer microoptics* [12]). The resulting potential for new methods of dispersion control may be an interesting task for future investigations. A presently more important step in the context of ultrashort pulses is the integration of materials with nonlinear optical characteristics (e.g., thin crystals or layers for second harmonic generation [13]) for the design of *nonlinear microoptical processors*, which enable higher-order autocorrelation data or intensity-dependent contrast enhancement. Recent results of theoretical and experimental work on the propagation of array-shaped free-space interference patterns (Talbot effect, arrays of Bessel-like beams [14]) generated with refractive, reflective, and hybrid optical microstructures and their application to the mapping of spatial and temporal coherence of femtosecond lasers are reported here.

2 Experimental Technique

Thin-film microoptical elements for femtosecond laser-beam shaping have been fabricated on the basis of vapor deposition technology and (for extremely small structure sizes) by gray-scale lithography. For vapor deposition, shading masks with single or multiple holes are fixed at substrates which undergo a planetary rotation and thus move on hypotrochoidal orbits relative to a pointlike source (electron-beam vaporizer) [4–6]. The technique enables the production of a variety of components, including multilayer microoptics [12], microlens arrays on flexible substrates [15], or anamorphotic microlenses [16]. The maximum height of the structures is limited by the deposition technology and depends on layer and substrate materials and the aspect ratio and does not exceed 15 µm (typically 1–6 µm for element diameters of 50–700 µm). The thickness or phase profile can be influenced by the mask parameters (distance, thickness, shape, depth profile). By choice of the material, broadband or selective transmission properties can be obtained. The transfer of deposited dielectric structures into the bulk materials by reactive ion etching extends the field of applications to other wavelength regions (e.g., down to far ultraviolet region, as recently demonstrated [17]).

For femtosecond laser-beam shaping and characterization, the following types of thin-film structures have been deposited and tested: (a) arrays of convex refractive silica microlenses on thin quartz substrates for a transmission setup, (b) gold-coated concave silica and copper structures as reflective shapers, and (c) hybrid refractive-reflective elements consisting of plane silver mirrors with a matrix of silica lenses on top. For the generation of Bessel-like beams, lenses and mirrors of Gaussian or inverse Gaussian phase profile were applied. Bessel beams were formed with conical type-B elements of straight slope. For self-imaging experiments with very small array periods, lithographically structured photoresist microlenses (d) were used (also transferred into the substrates or combined with reflecting metal layers).

In Figure 1, the thickness distribution of a part of a type-C hybrid microoptical beam shaper measured with a Mirau white-light interferometer (ZYGO) with $M = 5\times$ magnification at a wavelength of 616 nm is shown. The component consists of about 2000 convex silica axicons deposited on a silver mirror. To avoid parasitic Fabry–Perot interference effects by internal reflection, a part of the sample was coated with a thin auxiliary highly reflecting gold layer. The pitch of the hexagonal structure is 405 μm and the maximum structure height is about 1 μm. The Gaussian shape of the elements can be recognized in the cut-through structures in the first row of Figure 1.

To produce ultrashort pulses over a broad range of parameters, two Ti:sapphire laser systems with center wavelengths of about 790 nm were available. Sub-10-fs pulses were generated by a home-made amplified Ti:sapphire laser with a hollow fiber for self phase modulation, chirp-compensating prisms, and mirrors (pulse energy maximum 150 μJ, repetition rate 1 kHz [18]). From a second commercial system, oscillator pulses of about 10 fs (pulse energy 4 nJ, repetition rate 75 MHz) and amplified pulses of about 20 fs (repetition rate 1 kHz, pulse energy 800 μJ) were extracted. Polarization-maintaining beam guiding optical setups were used in all arrangements. The temporal and spectral characteristics of the pulses were determined by classical autocorrelation methods and a spectrometer. With symmetric (first system) and slightly asymmetric (second system) beam splitters of Mach–Zehnder and Michelson type, the laser beam could be divided into partial beams. For phase shifting, one of the arms was translated by a high-resolution piezoactuator (minimum step width 50 nm, corresponding to a time resolution of 0.34 fs). The interference patterns shaped by the microoptical components were imaged by an optical system consisting of a microscope objective ($M = 4\times$ or $20\times$) and a zoom objective (1:1.2/12.5–75 mm). The plane of interest was directly imaged to a charge-coupled device (CCD) matrix camera (Basler A101P, 1300×1030 pixels, pixel size 6.7×6.7 μm^2). Processing the pattern by second harmonic

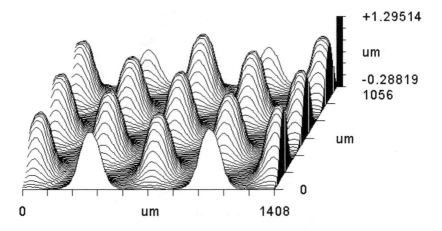

+1.29514

um

-0.28819
1056

um

0

0   um   1408

Fig. 1. Thickness distribution of a part of a type-C hybrid microoptical beam shaper (silica axicons on silver mirror, hexagonal structure, 405-µm pitch) measured with a Mirau white-light interferometer (ZYGO, magnification $M = 5\times$, wavelength 616 nm, auxiliary gold layer on a part of the sample).

generation (SHG) to obtain second-order autocorrelation was achieved by a BBO crystal (100 µm thickness) in the image plane. The fringes were analyzed with image processing software (Fringe Processor, BIAS). The position of the imaged plane was varied by shifting the BBO crystal, the objectives, and the camera simultaneously. The propagation was simulated on the basis of Rayleigh–Sommerfeld diffraction theory [14]. By the microoptical arrays, the laser beam is divided into separated subbeams for a further multichannel processing. At large distances, interference of diffracted and refracted parts appears (fractal Talbot effect, Talbot effect, far field) and the spectrum is split. In this region, the spatial and temporal resolution are reduced. The distance of optimum performance for Bessel beam generation depends on the layer thickness h and on the array period p (for a silica microaxicon array of $p = 405\,\mu\text{m}$ and $h = 5.7\,\mu\text{m}$, an optimum distance of 9–10 mm was found experimentally in good agreement with simulations [14, 19, 20]). At distances > 10 mm, the influence of diffraction is significant (the corresponding Fresnel number for the center wavelength $\lambda = 790$ nm was $p^2/4\lambda z = 5.2$).

3 Array Generation and Coherence Mapping

For any type of interferometer and beam shaper (focusing, array generation), the spatio-temporal characteristics of the transformed laser beam are closely connected to the coherence properties. Therefore, microoptical multichannel interferometry is of high practical relevance. In general, however, the retrieval of absolute coherence data from a time-integrated fringe pattern is not trivial. At ultrashort-pulse durations in the region of strong space-time coupling (about 10 fs for Ti:sapphire laser wavelengths), the spatial and temporal parameters of the wave packet are not separable. Therefore, the contrast contains information about both spatial and temporal correlations. For a qualitative evaluation of the adjustment state of a laser system or beam-shaping optics, a radially resolved contrast analysis can be very helpful. By transforming the beam into an array of interference patterns generated by corresponding subbeams, a discrete spatial resolution can be achieved. From the contrast data of all array elements, a *contrast map* can be extracted. For ultrashort-pulse lasers, the components for shaping of the interference patterns should not affect the temporal coherence by dispersion, nonlinear effects, or inhomogeneities. The spatial resolution depends on the method used and on the optical setup.

In the case of self-imaging, an array-shaped phase object of the period p is replicated in phase and amplitude in the jth Talbot plane at a distance $z = 2jp^2/\lambda$ for a monochromatic beam. The interference maxima contain information from j neighbors in each direction so that the minimum linear resolution is $2j$ times the array period ($\Delta x = 2jp$). For a broadband femtosecond beam, the contrast is reduced by the spectral spread in axial direction and by travel time differences [14] but allows a qualitative evaluation of the spatial homogeneity. The measured intensity distribution of a 12.5-fs Ti:sapphire laser beam in the first Talbot plane (related to the center wavelength of the spectrum of the fundamental wave) of a microlens array and the corresponding time-integrated contrast map are plotted in Figure 2. The array consists of an orthogonal arrangement of type-D resist structures (period 36 μm). The field of view (about $320 \times 320 \, \mu m^2$) represents the central region of the beam. The fine structure of the contrast map is indicative of a weak spatial inhomogeneity. If transversal information is encoded in the spatio-spectral beam parameters, Talbot processing may also be a way to read it out by spectrum-to-space conversion.

Coherence mapping with Bessel beam arrays delivers a higher spatial resolution (with respect to the array period p) and a lower sensitivity to axial position and tilt. If beam shapers with extremely small tilt angles of their optical faces are implemented (thin-film axicons), minimum travel time differences are obtained. Second-order fringe patterns (by second harmonic generation in a 100-μm BBO crystal) of a part of a hexagonal array of Bessel-like beams at a pulse duration of 8 fs (spectral coverage > 200 nm) generated with a hexagonal type-A microaxicon array (silica on quartz, maximum layer thickness 5.7 μm, period 405 μm) and the corresponding contrast map are depicted in Figure 3. To avoid saturation effects by overexposure, the contrast of an outer fringe (sixth maximum) was analyzed for each of the 23 sub beams (area $1.62 \times 1.62 \, mm^2$). In the central region (brighter zones), the coherence duration is shorter than in the outer region. This is indicated by a significantly reduced diameter of the fringe envelope in the center.

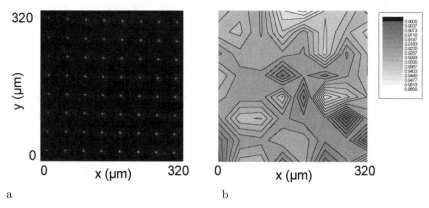

a   b

Fig. 2. Array shaping and coherence mapping by self-imaging of microlens arrays: (a) measured intensity distribution of a 12.5-fs Ti:sapphire laser beam in the first Talbot plane of the foci of an orthogonal arrangement of spherical resist microlenses (period 36 µm) and (b) corresponding contrast map indicating the fine structure of the coherence distribution in the beam center.

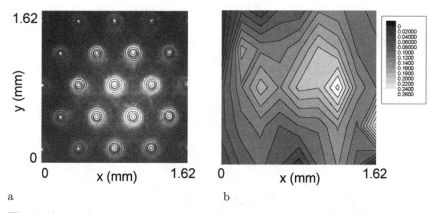

a   b

Fig. 3. Array shaping and coherence mapping with refractive thin-film microaxicon arrays: (a) measured second harmonic intensity distribution generated by Bessel-like subbeams at a distance of $z = 9$ mm (pulse duration 8 fs), and (b) corresponding contrast map indicating regions of shorter pulse duration in the beam center.

In both measurements, a transmittive setup was used to shape the beam of the fiber compressed laser system. To reduce the influence of the substrate by linear dispersion and nonlinear effects, reflective setups with type-C components can be used. The generation of spatially separated interference zones of high robustness against axial displacement and angular misalignment with the help of axicons can be applied to an extended beam characterization (including spatially resolved autocorrelators, wave-front sensing, and the detection of a spatial chirp) as well as to multichannel measuring techniques and materials processing.

4 Spatio-Temporal Autocorrelation of Bessel X-Pulses

The spectral interference of polychromatic nondiffracting waves results in characteristic spatio-temporal and spectro-temporal patterns, the so-called *X-waves* (X-shaped in space and time), which are a well-known phenomena in acoustics [21] and probably also have a close relationship to "monster waves" appearing at the ocean surface if waves of different wavelengths and directions interfere. In acoustics and in the radar technique, X-waves are used to generate directed energy for high-efficiency transducers or antenna systems. The vision of "electromagnetic missiles" was a further driving force in this field. With high-pressure lamps of short coherence time, continuous waves of X-shaped coherence distribution can be formed from conical beams [22]. At ultrashort-pulse durations with corresponding broad spectral bandwidths, pulsed X-waves or *X-pulses* are created. The launching of "light bullets" and their solitonlike propagation through dispersive media was predicted by theory [23]. Analyzing the propagation of conical 210-fs Ti:sapphire laser beams, it was concluded that optical X-pulses were shaped for the first time [24]. However, as could be shown by recent experiments [25] as well as by different numerical simulations, a splitting of the wave packet in distinct X-pulse structures of sufficiently high contrast can only be achieved at pulse durations < 20 fs [26–28]. In contrast to other experiments, we shaped and detected single and multiple Bessel-like and Bessel beams with conical elements of very small conical angles. In this case, the particular advantage of thin-film technology of enabling the fabrication of flat structures was exploited. In this way, ratios of depth of focus and minimum transversal diameter up to $2 \times 10^3 : 1$ were obtained. The space-time structure of the wave packets was analyzed by spatially resolved second-order autocorrelation (using second harmonic generation) for pulse durations between 8 and 100 fs. In Figure 4, the simulated evolution of a Bessel-like X-pulse during the propagation in vacuum in the axial direction is plotted for different distances (contrast of fine structures enhanced by modifying the luminance function).

The calculation is based on a single silica axicon microlens of Gaussian shape in transmission (height $5.7\,\mu$m, diameter $405\,\mu$m). For the spectrum, experimental data were used. It can be recognized that a localized wave packet appears, the spatial frequency of which varies slowly with the axial position. The region with best contrast is located between 8 and 10 mm from the vertex of the microaxicon. In this region, we also find maximum intensity and a minimum change of intensity along the axis. Therefore, a plane at $z = 9$ mm was imaged onto the CCD camera for the autocorrelation experiments. The second-order spatio-temporal autocorrelation of a selected microscopic-size subbeam of an array of nondiffracting beams is plotted in Figure 5. In these ex-

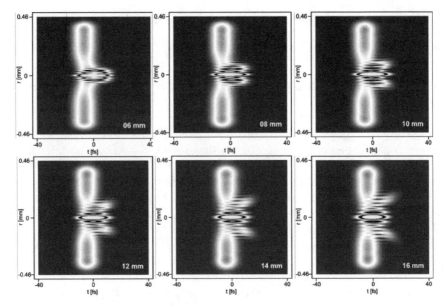

Fig. 4. Numerical simulation of the beam propagation: Spatio-temporal shape of a Bessel-like X-pulse at different distances related to the intensity (generating element: plano-convex silica axicon microlens of Gaussian shape in transmission, height 5.7 μm, period 405 μm, substrate directed to the laser, axicon directed to the opposite direction; parameter: distance from the vertex of the microaxicons to the center of the wave packets).

periments, we employed a Ti:sapphire laser of 12.5-fs pulse duration and used a type-A transmission setup. Figure 5 represents a two-dimensional, cross-sectional cut of a small part of the complete three-dimensional space-time autocorrelation. Because of the quadratic dependence of the second harmonic signal on the fundamental intensity, the intensity of the outer fringes falls with the inverse square of the radius or faster. To enhance the visibility of the X-shape, the gray values of the temporal profiles for all radial coordinates were normalized to their maxima. With these measurements, the first reliable and direct evidence of ultrashort-pulse localized wave packets of Bessel-X-pulse type at optical frequencies was demonstrated in good agreement with numerical simulations [26–28].

5 Wave-Front Autocorrelation with Advanced Shack–Hartmann Sensors

In addition to the contrast data, the analysis of array-shaped interference patterns also delivers information about the phase of the wave packet. A time-integrated measurement yields the local wave-front tilt averaged over all

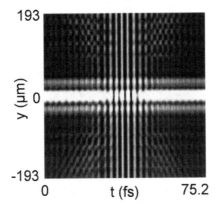

Fig. 5. Spatio-temporal second-order autocorrelation of localized ultrashort-pulse wave packets: Direct detection of a microscopic-size Bessel-like subbeam of an array of X-pulses (matrix of 200 beams generated from a 12.5-fs Ti:sapphire laser pulse, center wavelength 790 nm, microscope objective: $M = 4$, vertical axis: radial direction, horizontal axis: time, normalized to maximum intensity for each transverse coordinate y).

spectral contributions for each subaperture. Thus, a simultaneous measurement of wave front and autocorrelation function or *wave-front autocorrelation* is possible if the collinear autocorrelator setup is combined with a spatial separation in a sufficiently high number of channels by a microoptical array. Experiments were performed with transmitting type-A as well as with reflecting type-B and hybrid type-C setups. The thin-film components enable the extension of the Shack–Hartmann sensor principle to ultrashort pulses, whereas the use of axicons leads to extended foci and angular tolerance. In Figure 6, the second-order spatio-temporal autocorrelation of a femtosecond wave packet of an amplified Ti:sapphire laser is shown for two different radii of curvature by cuts through the three-dimensional distribution (two dimensions for the transversal profile, one dimension for the time axis, cut length 2.43 mm).

The direction of the cut was chosen in such a way that aberrations caused by the oblique illumination (here at an angle of incidence of 25°) can be neglected. The period of the undistorted beam (Fig. 6a) corresponds to the array period (405 µm), whereas the distorted beam had a local radius of curvature of about 38.2 mm (caused by a plano-convex glass lens of 2 diopters at a distance of 11.8 cm from the center of the array, thickness of the lens 2.05 mm, distance of the BBO crystal from the array 1.4 cm). In Figure 6b, reduced spacings of the subbeams clearly can be recognized (arrows). By measuring the average difference of the spacings, an averaged radius of curvature of the wave front of 37.5 cm was determined. This value agrees well with the parameters of the lens.

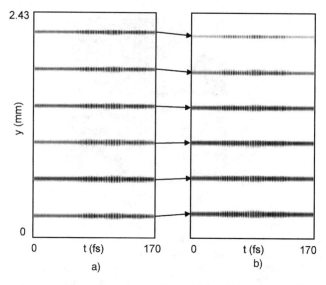

Fig. 6. Wave-front autocorrelation of second-order with hybrid refractive-reflective microaxicon arrays in reflective setup: (a) undistorted beam and (b) beam distorted by a thin glass lens (gray scale inverted, beam shaper: Gaussian-shaped silica axicon lenses on plane silver mirror, height of axicons about 1 μm, second harmonic generation: 100 μm BBO crystal, source: amplified Ti:sapphire laser, repetition frequency 1 kHz, central wavelength 790 nm, angle of incidence 25°, step width of piezotranslator 50 nm corresponding to a temporal resolution of 0.34 fs).

6 Summary and Conclusions

Thin-film microoptical components in refractive and hybrid refractive-reflective designs were applied to spatio-temporal shaping of ultrashort-pulse laser beams. Taking advantage of the specific properties of thin layers, the technique enables one to process pulses in the space-time coupling domain (sub-10-fs range) with reduced distortions. Arrays of conical beams of extremely small angles can be realized. The spatio-temporally resolved first- and second-order autocorrelation and the wave-front curvature have been measured simultaneously with advanced Shack–Hartmann sensors based on axicon arrays. Direct detection and characterization of single and multiple spatio-temporally localized Bessel X-pulses in the optical region was demonstrated for the first time. Reflection setups were realized with available reflective and hybrid refractive-reflective array components. For the development of future photonic systems, the integration of adressable and steerable structures (adaptive microoptics), the implementation of highly compact design (planar microoptics) and the incorporation of nonlinear materials (nonlinear microoptics) are of essential interest.

Acknowledgments

The authors acknowledge the contributions of H.-J. Kuehn (Quarterwave Berlin) to the development of deposition techniques. The success of the experiments was only possible thanks the close cooperation with U. Neumann and U. Griebner (MBI). We thank G. Steinmeyer, M. Tischer, W. Goleschny, M. Rini, A. Kummrow, E. T. J. Nibbering, K. Reimann, M. Woerner, G. Kordaß, R. Müller, and T. Elsasser (MBI), W. Jüptner (BIAS Bremen), M. Piché, G. Rousseau, N. McCarthy, M. Fortin, and S. L. Chin (Laval University Québec), E. Pawlowski (Heinrich Hertz Institute, Berlin), W. Seeber (Friedrich-Schiller-University, Jena), and K. Kolasinski (Queen Mary College, London), for support and exciting discussions. Visions and ideas were exchanged with P. Corkum (NRC Ottawa), P. Herman (University Toronto), S. Trillo (University of Ferrara), J. Jahns (FernUniversität Hagen), and A. Lohmann (University of Erlangen). The work was sponsored by DFG project GR 1782-2 and German Canadian Collaboration Project CAN 00/016 (BMBF/DLR).

References

1. Zewail, A.H., Femtochemistry: Atomic-scale dynamics of the chemical bond, J. Phys. Chem. A, 104, 5660–5694 (2000).
2. Konishi, T. and Ichioka, Y., Ultrafast temporal-spatial optical information processing, conversion, and transmission, in *Optical Information Processing. A Tribute to Adolf Lohmann*, edited by Caulfield, H.J., SPIE, Bellingham, WA (2002).
3. Trebino, R. and Zeek, E., Ultrashort laser pulses, in *Frequency-Resolved Optical Gating: The Measurement of Ultrashort Laser Pulses*, edited by Trebino, R., Kluwer Academic, Norwell, MA (2000).
4. Grunwald, R., Woggon, S., and Ehlert, R., Fabrication of thin-film microlens arrays by mask-shaded vacuum deposition, in *Diffractive Optics and Optical Microsystems*, Martellucci, S. and Chester, A.N. (eds.), Plenum Press, New York (1997).
5. Grunwald, R., Mischke, H., and Rehak, W., Microlens formation by thin-film deposition with mesh-shaped masks, Appl. Opt. 38, 4117–4124 (1999).
6. Grunwald, R., Mikrooptische Komponenten für neue Technologien, in *Das Handbuch der Bildverarbeitung*, edited by Ahlers, R.-J., Expert-Verlag, Malmsheim (2000).
7. McLeod, J.H., The axicon: a new type of optical element, J. Opt. Soc. Am. A 44, 592–597 (1954).
8. Grunwald, R., Woggon, S., Griebner, U., Ehlert, R., and Reinecke, W., Axial beam shaping with non-spherical microlenses. Jpn. J. Appl. Phys., 37, 3701–3707 (1998).
9. Grunwald, R., Woggon, S., Ehlert, R., and Reinecke, W., Thin-film microlens arrays with non-spherical elements, Pure Appl. Opt., 6, 663–671 (1997).
10. Durnin, J.E., Exact solutions for nondiffracting beams. I. The scalar theory, J. Opt. Soc. Am. A 4, 651–654 (1987).

11. Herman, R.M. and Wiggins, T.A., Production and uses of diffractionless beams, J. Opt. Soc. Am. A 8, 932–942 (1991).

12. Grunwald, R., Nerreter, S., Griebner, U., and Kuehn, H.-J., Design, characterization and application of multilayer micro-optics, Proc. SPIE 4437, 40–49 (2001).

13. Neumann, U., Grunwald, R., Griebner, U., Steinmeyer, G., Woerner, M., and Seeber, W., Second harmonic characteristics of photonic composite glass layers with ZnO nanocrystallites for ultrafast applications, Proc. SPIE 4972, 112–121 (2003).

14. Grunwald, R., Griebner, U., Elsaesser, T., Kebbel, V., Hartmann, H.-J., and Jüptner, W., Femtosecond interference experiments with thin-film micro-optical components, in *Fringe 2001*, edited by Osten, W. and Jüptner, W., Elsevier, Paris (2001).

15. Grunwald, R., Ehlert, R., Woggon, S., and Witzmann, H.-H., Thin-film microlens arrays on flexible polymer substrates, in *Micro System Technologies*, edited by Reichl, H., and Heuberger, A., VDE-Verlag, Berlin 793–795 (1996).

16. Griebner, U. and Grunwald, R., Generation of thin film microoptics by crossed deposition through wire grid masks, Proc. SPIE 3825, 136–143 (1999).

17. Grunwald, R., Neumann, U., Kebbel, V., and Mann, K., Thin-film microaxicon arrays for beam shaping and characterization of vacuum ultraviolet lasers, CLEO Europe, 2003, June 22–27, Munich, Conference Digest (2003).

18. Nibbering, E.T.J., Dühr, O., and Korn, G., Generation of intense tunable 20-fs pulses near 400 nm using a gas-filled hollow waveguide, Opt. Lett. 22, 1335–1337 (1997).

19. Grunwald, R., Griebner, U., Tschirschwitz, F., Nibbering, E.T.J., Elsaesser, T., Kebbel, V., Hartmann, H.-J., and Jüptner, W., Generation of femtosecond Bessel beams with microaxicon arrays, Opt. Lett. 25, 981–983 (2000).

20. Grunwald, R., Griebner, U., Nibbering, E.T.J., Kummrow, A., Rini, M., Elsaesser, T., Kebbel, V., Hartmann, H.-J., and Jüptner, W., Spatially resolved small-angle non-collinear interferometric autocorrelation of ultrashort pulses with microaxicon arrays, J. Opt. Soc. Am. A 18, 2923–2931 (2001).

21. Lu, J.Y. and Greenleaf, J.F., Nondiffracting X-waves. Exact solutions to free space scalar wave equation and their finite aperture realizations, IEEE Transactions on Ultrasonics, Ferroelectrics, and Frequency Control 39, 19–31 (1992).

22. Saari, P. and Reivelt, K., Evidence of x-shaped propagation-invariant localized light waves, Phys. Rev. Lett. 79, 4135–4138 (1997).

23. Conti, C., diTrapani, P., Trillo, S., Valiulis, G., Minardi, S., and Jedrkiewicz, O., Nonlinear X-waves: light bullets in normally dispersive media? QELS 2002, Long Beach, Technical Digest, 112–113 (2002).

24. Sonajalg, H., Rätsep, M., and Saari, P., Demonstration of the Bessel-X-pulse propagating with strong lateral and longitudinal localization in a dispersive medium, Opt. Lett. 22, 310–312 (1997).

25. Piché, M., Rousseau, G., Varin, Ch., and McCarthy, N., Conical wave packets: their propagation speed and their longitudinal fields, Proc. SPIE 3611, 332–343 (1999).

26. Grunwald, R., Griebner, U., Neumann, U., Kummrow, A., Nibbering, E.T.J., Rini, M., Kebbel, V., Piché, M., Rousseau, G., and Fortin, M., Femtosecond laser beam shaping with structured thin-film elements, Proc. SPIE 4833, 354–361 (2002).

27. Grundwald, R., Griebner, U., Neumann, U., Kummrow, A., Nibbering, E.T.J., Piché, M., Rousseau, G., Fortin, M., and Kebbel, V., Generation of ultrashort-pulse nondiffracting beams and X-waves with thin-film axicons, in *Ultrafast Phenomena XIII*, edited by Miller, R.D., Murnane, M.M., Scherer, N.F., and Weiner, A.M., Springer-Verlag New York (2002).
28. Grunwald, R., Kebbel, V., Griebner, U., Neumann, U., Kummrow, A., Rini, M., Nibbering, E.T.J., Piché, M., Rousseau, G., and Fortin, M., Generation and characterization of spatially and temporally localized few-cycle optical wave packets, Phys. Rev. A 67, 063820 (2003).

Index

Springer Series in
OPTICAL SCIENCES

Springer Series in
OPTICAL SCIENCES